REVISING THE TWO MTW FORCE SHAPING PARADIGM

A "Strategic Alternatives Report"
From the Strategic Studies Institute

Edited by
Steven Metz

April 2001

*****

*****

Comments pertaining to this report are invited and should be forwarded to: Director, Strategic Studies Institute, U.S. Army War College, 122 Forbes Ave., Carlisle, PA 17013-5244. Comments also may be conveyed directly to the editor. Dr. Metz can be contacted at (717) 245-3822, fax (717) 245-3820, or email at Steven.Metz@carlisle.army.mil. Copies of this report may be obtained from the Publications and Production Office by calling commercial (717) 245-4133, FAX (717) 245-3820, or via the Internet at Rita.Rummel@carlisle.army.mil

*****

Most 1993, 1994, and all later Strategic Studies Institute (SSI) monographs are available on the SSI Homepage for electronic dissemination. SSI's Homepage address is: http://carlisle-www.army.mil/usassi/welcome.htm

*****

The Strategic Studies Institute publishes a monthly e-mail newsletter to update the national security community on the research of our analysts, recent and forthcoming publications, and upcoming conferences sponsored by the Institute. Each newsletter also provides a strategic commentary by one of our research analysts. If you are interested in receiving this newsletter, please let us know by e-mail at outreach@awc.carlisle.army.mil or by calling (717) 245-3133.

ISBN 1-58487-049-4

CONTENTS

FOREWORD

U.S. military strategy is undergoing its most serious examination since the end of the Cold War. Led by Secretary of Defense Donald Rumsfeld, this process is designed to assess every dimension of the strategy, including its most basic assumptions and concepts. For the first time in over a decade, everything about U.S. military strategy is subject to question.

One of the most important elements of U.S. military strategy for the past ten years has been the belief that a force able to fight two nearly simultaneous major theater wars (MTW) of the DESERT STORM type would be capable of dealing with the full gamut of security challenges that the United States is likely to face. Now nearly every expert on U.S. military strategy agrees that this force shaping paradigm needs a relook.

Beyond that point of agreement, the experts diverge. Some contend that DESERT STORM-type wars are so unlikely that creating a military focused on them is a waste of money. But even this group does not agree on what should replace the two MTW force shaping paradigm. Other experts are less willing to abandon the paradigm, arguing instead that it simply needs updating to reflect the nature of the contemporary security environment.

In the Strategic Alternatives Report that follows, the Strategic Studies Institute has collected essays from some of the leading thinkers on the question of a force shaping paradigm for the U.S. military. They represent a wide range of scholars, analysts, government officials, and uniformed thinkers. While they share a belief in the importance of the issue, they vary widely on their assumptions, analytical parameters, and recommendations. By airing this range of alternative perspectives, the Strategic

Studies Institute hopes to contribute to the building of a new, 21st century U.S. military strategy.

DOUGLAS C. LOVELACE, JR
Director
Strategic Studies Institute

CHAPTER 1

INTRODUCTION

Steven Metz

Since the early 1990s, U.S. military strategy has called for a force able to fight and win two nearly-simultaneous major theater wars (MTWs). An MTW was something similar to Operation DESERT STORM—a large-scale conventional war in Eurasia against an aggressive regional power involving substantial American forces from all services and, most likely, allies or coalition partners. While policymakers and planners admitted that the outbreak of two nearly simultaneous MTWs was unlikely, they felt that a military able to deal with such a challenge would also be sufficient for other likely missions and tasks.

The two MTW force shaping paradigm had a tremendous effect on American strategy. National security strategy always has both internal and external dimensions. The external dimension specifies how a nation will deal with others, specifically how it will use power resources to protect or advance its interests. The internal dimension deals with building a consensus on strategic objectives and methods, mobilizing support for the strategy, and developing the means to attain the objectives, including a military force. Many factors shape force development, including geography, wealth, strategic culture (including a nation's tolerance for risk, its aggressiveness, and its worldview), level of technological development, strategic objectives, the shape and form of threat, domestic politics, commitments, partnerships, and expectations concerning the nature of current and future armed conflict.

These factors made the two nearly-simultaneous MTW force shaping paradigm the appropriate one in the 1990s.

Geography dictated that any war involving the U.S. military would take place overseas. Only Eurasia combined a high enough level of U.S. national interest and of threat to lead the United States into a large scale war. The wealth of the United States meant that the nation could support a force that was large enough and advanced enough to project decisive power to nearly every part of the globe. Americans saw armed conflict as abnormal and episodic, arising from aggression by dictators. Such dictators, in the American worldview, only understood force, and thus must be deterred or defeated rather than accommodated. If one succeeded at armed aggression, others would be inspired to attempt it. This meant that the United States sought to deter or reverse aggression in a wide band where vital interests were at stake, including Europe, Southwest Asia, and Northeast Asia.

The best way to deter aggression, American strategic thinkers held, was through military preponderance. During the 1990s, the assumption was that the primary threat to regional stability and U.S. interests came from "rogue states" largely armed with Soviet or Soviet-style equipment, and thus preferring to fight a Soviet-style war based on armor-heavy conventional units. Such rogue states wanted to expand their territory or augment their national power by controlling vital resources and were willing to use force to do so. Because the primary threat—defined as one that was both possible and potentially dangerous—came from rogue states, the United States could expect to fight with allies or coalition partners in an MTW.

Continuing a trend that began after World War II, American military strategy in the 1990s was based on the qualitative superiority of U.S. forces, both in technology and equipment, and in human factors like training and leadership. Qualitative superiority, along with alliances and partnerships, allowed a relatively small U.S. military to undertake global commitments. If war against a major opponent did occur, qualitative superiority could help compensate for numerical inferiority, particularly early in a war before all American and allied units had deployed. America sought a quick resolution of armed conflict since short wars would be more likely to be supported by the American people and their elected leaders. The two MTW force shaping construct, in other words, had both a quantitative dimension and a temporal one. A military built with it in mind could not only win two wars nearly simultaneously, but could do so relatively quickly.

The nature of the global security environment of the 1990s and of U.S. national security policy made the two MTW force shaping paradigm logical. But does this still hold? In some fundamental ways, the global security environment of 2001 differs from earlier ones. For instance, while the degree of danger in the system has not diminished, the nature of the threat has. Cross-border invasion by a rogue state with a Soviet-style military is much less likely today than a decade ago. Humanitarian intervention and protracted peacekeeping are increasingly important tasks for the militaries of many nations, including the United States. Globalization and the information revolution have linked the world more closely than at any time in history, thus making it more difficult to ignore aggression or instability in less vital regions. And the ongoing revolution in military affairs, in combination with the accelerated pace of change which characterizes the 21st century, has led American policymakers and planners to conclude that they must undertake a fundamental transformation of the U.S. military. To build a military that will remain dominant twenty years in the future at the same time that it is able to protect current regional stability is a daunting task.

These changes in the global security environment have caused many American defense thinkers, both in and out of the government, to question the two MTW force shaping paradigm. That it needs reassessed is clear. But beyond that, questions abound. Does the two MTW force shaping paradigm need updated or replaced? If American strategists replaced it, what should the new paradigm be? If revision is enough, what should be changed? Is it enough to simply redefine "major theater war" or should there be a different building block? Should conventional wars in Southwest or Northeast Asia continue to serve as the model for the concept of a major theater war? Is a force that can deal with two nearly simultaneous MTWs truly flexible enough to also meet other security threats that are both possible and potentially dangerous?

The time is right to ask these questions. This is likely to be an extremely important year in the evolution of American national security and military strategy. As a new administration assumes power, the Pentagon is undertaking a congressionally-mandated Quadrennial Defense Review. The questions associated with the two MTW forcing shaping paradigm will be a vital part of the strategy formulation process of the new administration, and of the Quadrennial Defense Review. While the final answers to them are not year clear, a range of possible answers has emerged. The essays in this collection delineate a range of approaches to force shaping issues. While they are unified by the strategic acumen of their writers and by the fact that they all address the same key questions, they do not all share assumptions or, more importantly, conclusions. They recommend solutions ranging from a relatively modest updating of the two MTW forcing shaping paradigm to its abandonment.

The usefulness and validity of any force shaping paradigm must be assessed using a range of variables. One of the simplest yet most powerful ways of doing this is to analyze the suitability, feasibility, and acceptability of each alternative or recommendation. The analysis of suitability asks whether an alternative is likely to attain its objectives. The suitability of a force shaping paradigm must be assessed in terms of acceptable risk, national interests, and the nature of the threat or challenges that might be faced. A suitable alternative is one with a good chance of promoting or protecting national interests and deterring or defending against threats at an acceptable level of risk. In order to meet the suitability test, all of the force shaping paradigms discussed in this collection make certain assumptions about future U.S. national interests, future threats or challenges, and the level of acceptable risk. If these assumptions do not hold, the paradigm is not suitable. The feasibility of a force shaping paradigm is based on the existence of adequate resources. Each paradigm is based on assumptions about future defense budgets; demographics; the ability of the U.S. military to recruit soldiers, sailors, airmen, and Marines; the contribution of allies or partners; and technology. Again, the validity of a paradigm is no greater than the validity of its assumptions. Finally, the acceptability of a force shaping paradigm is contingent on its being supported by the American people and their elected leaders. A force shaping paradigm which would generate a military or a military strategy that is unlikely to garner this support—for instance, one based on the preemptive use of weapons of mass destruction—would not pass the acceptability test, and therefore has little value.

As will become clear, the force shaping paradigms offered in this volume differ as to their assumptions and on the relative priority assigned to suitability, feasibility, and acceptability. Some are imminently suitable, but may not be adequately feasible and acceptable. Some are feasible, but make not assure attainment of national objectives. Strategy always entails difficult trade-offs and the setting of priorities. That holds as much for the construction of a force shaping paradigm as for any other element of strategy. Somewhere within the alternatives presented here, though, lies a force shaping paradigm that maximizes suitability, feasibility, and acceptability. It may be one of those proposed by the authors, or it may be a synthesis of several of them. This can only become clear through analysis, debate, and consensus building. By illuminating the key components and assumptions of the force shaping problem and providing a range of cogent and creative solutions, these essays make a major contribution to this process.

CHAPTER 2

SIZING THE FORCE FOR THE 21st CENTURY

John F. Troxell

You cannot make decisions simply by asking yourself whether something might be nice to have. You have to make a judgement on how much is enough.

Robert S. McNamara
April 20, 1963

Ever since the end of the Cold War, the United States has been struggling to gain consensus on an appropriate force planning methodology concerning the size of its military establishment and to answer the question "how much is enough?" The size and posture of the U.S. military was the principal topic of the first Quadrennial Defense Review (QDR) and the National Defense Panel's (NDP) Alternative Force Structure Assessment, and remains an important task for the U.S. Commission on National Security/21st Century. The National Defense Authorization Act for fiscal year 2000, which made QDRs a permanent requirement, included under its first and principal task to the Secretary of Defense a call for a comprehensive discussion of national defense strategy and the force structure best suited to implement that strategy. Consequently, the new administration, and all subsequent "new administrations" must explicitly show their hand at the complex task of force planning.

In designing forces to protect U.S. national interests, military planners must accomplish three tasks: determine how much force is required to protect those interests with a certain degree of assured success or a minimum degree of acceptable risk; determine how to posture that force; and finally convince Congress and the public that the solutions for the first two tasks are reasonably correct.[1] The issue of creating well-reasoned force structure requirements and convincing cost conscious politicians is not an inconsequential matter.

Most defense analysts claim that during the Cold War the force planning task was relatively straightforward. The threat posed by the Soviet Union required the fielding of forces capable of conducting a global war, with priority placed on defending Western Europe. This situation served as the agreed scenario around which to design and develop forces and measure risks if specific force goals were not met. Force modernization programs were also directly linked to maintaining a qualitative advantage over projected improvements in Soviet capabilities. The resultant Cold War force was so large that all other military requirements, such as forces for forward presence, smaller-scale interventions, and humanitarian operations, could be met as lesser-included requirements.

During the post-Cold War period, the sizing function that replaced the global war scenario against the Soviet Union has been the requirement to be able to prosecute major theater war (MTW). This requirement evolved during the last years of the Bush administration as the rationale for the Base Force. The first act of the new Clinton administration was to reexamine the issue, resulting in the Bottom Up Review (BUR) Force. The Base Force and the BUR Force were both sized against the requirement to fight two MTWs. This force-sizing requirement was revalidated in the 1997 Quadrennial Defense Review, but continues to generate a great deal of controversy. Depending on the point of view, the force structure associated with this posture is attacked for being over-stuffed, unaffordable, or totally inadequate.[2] During the past 15 months a working group at the National Defense University has been examining issues for the 2001 QDR and recently concluded that a severe strategy-resources gap exists and that the 2001 QDR must confront the "iron triangle" of fundamental choices—spend more, cut costs, or do less.[3] The two MTW force-sizing construct is endemic to all of these choices. The purpose of this chapter is to clarify the role of the two MTW requirement within the current defense program and propose an alternative force-sizing construct.

Force Planning Methodologies.

Since the advent of the Cold War, military planners have used two very different force-planning methodologies.[4] The easiest to conceptualize is threat-based planning. This methodology is preeminent when threats to U.S. interests are easily recognized and identified. The task for the planner is to postulate a reasonable scenario, or a specific military contingency, then determine the amount of force needed to prevail in that scenario. This approach lends itself to dynamic and static modeling and provides a quantifiable rationale for the recommended force structure, and answers the question: Can the United States, either unilaterally or as part of an alliance, defeat the opponent or prevail in the postulated contingencies? The logic of this approach is very compelling and greatly facilitates accomplishing the planner's third task—convincing the public and Congress.

The second major methodology is generally referred to as capabilities-based planning. Somewhat harder to conceptualize, analysts have proposed several variants of the same basic theme. Capabilities-based planning is most in vogue when threats to U.S. interests are multifaceted and uncertain, and do not lend themselves to single point scenario-based analysis. Instead of focusing on one or more specific opponents, the planner applies a liberal dose of military judgment to determine the appropriate mix of required military capabilities. Capabilities-based planners claim to focus on objectives rather than scenarios. A major problem planners have with this approach is convincing Congress that military judgment has established the proper linkage between the recommended force and the uncertain geostrategic environment.

POST-COLD WAR FORCE PLANNING

"Uncertainty is not a mere nuisance requiring a bit of sensitivity analysis," Paul Davis points out; "it is a dominant characteristic of serious planning."[5] The U.S. military is well aware of this fact, but has had difficulty during the current strategic transition in selling it to Congress and the public. The principal problem is the lack of the all-consuming threat that focused the nation's attention on the problem of containing the USSR for over four decades. The global war force planning framework has evaporated in the post-Cold War era, leaving little agreement on appropriate threats, contingencies, or required capabilities against which to focus the defense establishment.

The Base Force.

In an effort to demonstrate military responsiveness to changes in the strategic and budgetary environments, the Chairman of the Joint Chiefs of Staff, General Colin Powell, developed the Base Force in the early 1990s. This force was considered the minimum force that would still allow the armed forces to meet mission requirements with acceptable risk. The Base Force straddled both the Soviet revolutions of 1988 and 1991, causing the justification and rationale behind the chosen force levels to evolve over time. The initial focus of the Base Force was on a capabilities-based approach to defense planning, driven largely by resource constraints.

The threat was very ill-defined at this point. "I'm running out of demons," General Powell commented in April 1991, "I'm running out of villains. . . . I'm down to Castro and Kim Il Sung."[6] In such an environment, Powell stressed there were some very real limitations to threat-oriented contingency analysis. The resource-constrained force, he concluded, should instead focus on the combat capabilities needed to ensure that a sufficient array of assets would be present to perform the multiple missions demanded on the modern battlefield.[7] The mission-focused aspect of the Base Force was evident in the three conceptual conventional force packages that eventually became part of the 1992 National Military Strategy (NMS). Forces for the Atlantic would include forward-based and forward deployed units committed to Europe, and heavy reinforcing forces for Europe, the Middle East, and the Persian Gulf based in the United States. The Pacific Forces differed from the Atlantic package, reflecting the maritime character of the area. Finally, Contingency Forces would consist of U.S. based ground, air, and naval forces capable of worldwide deployment as needed.[8]

Unfortunately, the advent of and ensuing focus on Operations DESERT SHIELD and DESERT STORM precluded the Pentagon's strategic planners from completing the analytical construct behind the Base Force, a task that Congressman Les Aspin was more than willing to undertake. In the first of two national security papers, Aspin attacked capabilities-based force planning, charging that decisions concerning what capabilities were required of U.S. forces could not be done in a vacuum. Instead, he concluded, ". . . it is critical to identify threats to U.S. interests that are sufficiently important that Americans would consider the use of force to secure them."[9] Shortly thereafter, Aspin outlined in a second paper his concept of the "Iraqi equivalent" as the generic threat measure for regional aggressors and the "Desert Storm equivalent" as the most robust building block for U.S. forces. The purpose was to establish a clear linkage between the force structure and the sorts of threats the forces could be expected to deal with. Aspin also envisioned his "threat-driven" methodology to be flexible enough to include aspects of a typical capabilities-based approach. The building blocks for the methodology, he pointed out, were generic capabilities.

> Although each is informed by a careful review of pertinent historical cases, I am not suggesting we acquire forces which would be suited only to a few places and precedents. I'm suggesting instead generic military capabilities which should be effective against the full spectrum of categorical threats in the uncertain future.[10]

Partly in response to this criticism the rationale for the Base Force evolved into a combined capabilities-based and threat-based approach and became more firmly anchored to the two MTW requirement. In late 1992, General Powell began promoting the Base Force as both capabilities oriented as well as threat oriented. In a few cases such as Korea and Southwest Asia, he pointed out, it was possible to identify particular threats with some degree of certainty.[11] These developments had no effect on the regional focus of the force. In 1992, Secretary of Defense Richard Cheney reported that, "the ability to respond to regional and local crises is a key element of our new strategy."[12] The "Base Force" National Military Strategy of 1992 concluded that U.S. "plans and resources are primarily focused on deterring and fighting regional rather than global wars."[13] Although neither of these documents specified a two MTW requirement, the sizing function for this requirement continued to evolve behind the scenes. Both the 1991 and 1992 Joint Military Net Assessments (JMNAs) focused on the warfighting analysis for Major Regional Contingency-East (MRC-East)-Southwest Asia, and MRC-West-Korea. According to Army force planners, the principal focus of U.S. operational planning was "regional crisis response—to include a capability to respond to multiple concurrent major regional contingencies."[14] In his autobiography General Powell clearly states what his National Military Strategy did not: "The Base Force strategy called for armed forces capable of fighting two major regional conflicts 'nearly simultaneously.'"[15]

The Bottom Up Review Force.

With a new administration, the Base Force title was jettisoned; but the underpinnings of U.S. force structure remained largely intact. Upon assuming office, Secretary of Defense Les Aspin initiated a comprehensive review of the nation's defense strategy and force structure and published the *Report of the Bottom Up Review (BUR)* in October 1993. The methodology for the *BUR* combined all threat-based and capabilities-based aspects of the force-planning methodologies. To begin with, there was the traditional assessment of threats and opportunities, the formulation of a strategy to protect and advance U.S. interests, and the determination of the forces needed to implement the strategy. At the same time, there was an evaluation of military missions that included fighting MTWs, conducting smaller-scale operations, maintaining overseas presence, and deterring attacks with weapons of mass destruction. The ultimate force-sizing criterion was to "maintain sufficient military power to be able to win two major regional conflicts that occur nearly simultaneously." The planning and assessment for these MTWs were based on two illustrative scenarios viewed as representative yardsticks with which to assess in "gross terms the capabilities of U.S. forces."[16] From this perspective, the *BUR* continued the dual focus on both threat and capabilities that had evolved in the Base Force. "The Clinton defense policy," noted defense analyst Richard L. Kugler points out,

> represents continuity rather than a revolutionary departure, for the changes it makes are relatively small. . . . The chief difference lies in the new policy's call for a smaller conventional posture, but only 10-15 percent smaller than the Bush administration's Base Force.[17]

Quadrennial Defense Review (QDR) and the National Defense Panel (NDP).

Despite a degree of continuity and general agreement within the nation's defense establishment concerning the overall framework for the size and posture of U.S. military forces, planners continued to have difficulty with their third task—convincing Congress and the public. The greatest difficulty was persuading Congress that the Pentagon was sufficiently focused on the 21st century and preparing the military to execute the most likely conflicts. As a result, in 1996 Congress passed legislation directing the Secretary of Defense and the Chairman of the Joint Chiefs of Staff to conduct a review of the U.S. defense program and provide a report in 1997. Their review was directed to include "a comprehensive examination of the defense strategy, force structure, force modernization plans, infrastructure, budget plan and other elements of the defense program . . ."[18] Congress also provided for an independent body of defense experts, designated the National Defense Panel, to both review and comment on the *QDR*, as well as look slightly further into the future.

The *QDR* was designed as a strategy-driven review and upon its completion to serve as the overall strategic planning document for the Defense Department. From a force planner's perspective the key features of the *QDR* were the newly articulated defense strategy of "shape, respond, and prepare," and several refinements to force sizing and planning considerations. However, the bottom-line remained an overall requirement "that U.S. forces must be capable of fighting and winning two major theater wars nearly simultaneously."[19]

The shape-respond-prepare strategy recognized the requirements for U.S. military forces to operate in support of U.S. interests across the entire spectrum of operations, from peacetime to wartime. Military forces assist in shaping the international environment through overseas presence, rotational deployments, and various military-to-military programs. Shaping requirements have normally been viewed as a lesser-included capability provided by a larger war-time focused structure. The *QDR*, however, specifically indicated that the overseas presence mission plays a significant role in determining the size of U.S. naval forces.[20] Responding to the full spectrum of crises, to include major theater wars, remained the most stressing requirement. Although the *QDR* revalidated the centrality of a 2 MTW force structure, it also placed increased emphasis on capabilities needed for smaller-scale contingencies (SSC). These contingencies are viewed as the most likely challenge for U.S. forces, and the *QDR* noted a requirement to be able to conduct multiple concurrent smaller-scale contingency operations.[21] One of the difficulties in using this approach as a force structure determinant, however, is that while the military is relatively confident that it knows the types and quantity of forces needed to fight an MTW, it is much less certain of what is needed for SSCs that have a wide variety of objectives and occur in diverse regions of the world.[22] The Joint Staff sponsored a wargame series known as Dynamic Commitment that attempted to identify and quantify a list of military capabilities for smaller-scale contingencies. These capabilities, however, are still viewed as a lesser-included subset of the MTW force.

The *QDR's* analysis continued to represent a blend of threat-based and capabilities-based planning. The principal scenarios remain focused on the threat posed by regional aggressors on the scale of Iraq or North Korea. A slightly expanded scenario set was used to examine threat use of asymmetric strategies, differences in warning time, U.S. force size, and the degree of commitment to ongoing SSCs. The *QDR* also tested projected capabilities against a range of more challenging threats—a postulated major regional power in the 2014 timeframe. In addition, generic scenarios used a threat force based on the projected capabilities of nations not currently allied with the United States. As the report concludes, "this analysis enabled us to test our projected capabilities against a range of more challenging threats."[23]

The report of the National Defense Panel highlighted another dilemma faced by force planners—building forces for the present or focusing on future requirements. Concerning the present, the NDP acknowledged that the United States cannot afford to ignore near-term threats and that "the two-theater construct has been a useful mechanism for determining what forces to retain as the Cold War came to a close, [and] to some degree, it remains a useful mechanism today."[24] But the panel also argued that today's threats are not necessarily the ones the United States will face in the future, expressing concern that the two MTW construct is becoming an inhibitor to achieving the capabilities needed in the 2010-2020 time frame. The panel suggested a fundamental change: "The United States needs a transformation strategy that enables us to meet a range of security challenges in 2010-2020 without taking undue risk in the interim."[25]

Two MTW Rationale.

In examining the rationale for the two MTW requirement, it is important to remember that the requirement was never construed to be a strategy, but represented the sizing function for the Clinton administration's defense program—the principal determinant of the size and composition of U.S. conventional forces. The nature of this sizing function was clearly articulated by former Defense Secretary William Perry in 1996:

> Previously, our force structure was planned to deter a global war with the Soviet Union, which we considered a threat to our very survival as a nation. All other threats, including regional threats, were considered lesser-but-included cases . . . Today, the threat of global conflict is greatly diminished, but the danger of regional conflict is neither lesser nor included and

> has therefore required us to take this danger explicitly into account in structuring our forces.[26]

The current version of the two MTW requirement states that the principal determinant of the size and composition of United States conventional forces is the capability "preferably in concert with allies, . . . to deter and, if deterrence fails, defeat large-scale, cross-border aggression in two distant theaters in overlapping timeframes."[27] Inherent in the acceptance of the two MTW force-sizing requirement was the recognition that the United States would not be able to conduct sizable contingency operations at the same time it was fighting in two major theaters.[28]

Three principal reasons for this sizing function emerged during the post-Cold War period. First, as a nation with global interests, the United States needed to field a military capability to credibly deter opportunism—avoiding a situation in which it lacked the forces to deter aggression in one region while fighting in another. "A one-theater war capacity," the *QDR* points out, "would risk undermining both deterrence and the credibility of U.S. security commitments in key regions of the world."[29] The vitality of U.S. alliance relationships is critical to the nation's security. Insufficient and thus incredible military capability to respond and protect worldwide interests could severely undermine that vitality. The historical evidence in support of the two MTW requirement is much stronger than detractors are willing to acknowledge. There have been, for instance, 22 nearly simultaneous crises requiring the deployment and use of military force from 1946 to 1991.[30] It is argued that the likelihood of concurrent crises has increased in the absence of the Cold War superpower restraints.

A second reason was that a force capable of defeating two regional adversaries should provide the basic wherewithal to support a defense against a larger-than-expected threat or respond to a regional crisis under circumstances more difficult than expected.[31] Although a peer competitor is still not envisioned in the near term, the possibility of confrontations with a larger than MTW threat must be guarded against. This hedge against uncertainty was also required as a practical matter because of the time needed to reconstitute a larger force. "If we were to discard half of this two MTW capability or allow it to decay," the former Chairman of the Joint Chiefs of Staff, General Shalikashvili, concluded a few years back,

> it would take many years to rebuild a force of comparable excellence. In today's turbulent international environment, where the future posture of so many powerful nations remains precarious, we could find ourselves with too little, too late.[32]

Finally, the two MTW sizing function recognized the increased operational deployment of American forces and provided adequate force to allow the United States to deter latent threats from regional adversaries when portions of the force were committed to important smaller-scale contingencies and engagement activities in other theaters.[33] Although U.S. participation in smaller-scale contingency operations should not be viewed as a given, if the National Command Authorities (NCA) decided to commit U.S. forces to such operations, the strategy and force structure, as sized by the two MTW requirement, was thought to adequately support that commitment. One of the current points of contention is the argument that the two MTW force lacks appropriate capabilities and possibly the size to support rotational policies associated with the increased scope of SSCs.

Components of the Two MTW Force-sizing Construct.

In addition to the rationale discussed above, the current defense program, unlike its Cold War predecessors, provides a great deal of specificity concerning important operational components of the 2 MTW sizing function. The first component is the two illustrative planning scenarios (IPS) developed to assist planning and assessment. The IPS depict aggression by a remilitarized Iraq against Kuwait and Saudi Arabia, and by North Korea against the Republic of Korea. Each scenario examines the performance of projected U.S. forces in relation to critical parameters, including warning time, threat, terrain, regional allies, and duration of hostilities.[34] These scenarios were not designed to replicate the operational plans of the warfighting CINCs, but rather to assess forces and support assets for a wide range of possible future operations.[35] In addition to the MTW scenarios, the defense program has also examined numerous smaller-scale operations in order to identify any unique force requirements not specified in the 2 MTW warfight.

A second component is a notional operational scheme for the execution of an MTW. U.S. planning for fighting and winning MTWs envisions an operational strategy that in general unfolds as follows:

- halt the invasion;
- build-up U.S. and allied combat power in theater while reducing the enemy's;
- decisively defeat the enemy;
- provide for post-war stability.[36]

The final component is the MTW building block. According to the 1996 DOD Annual Report, "the following forces will be adequate, under most conditions, to successfully fight and win a single MTW," assuming continued progress on programmed force enhancements to strategic lift, prepositioning, and other force capabilities and their support assets:

- 5 Army divisions;
- 10 Air Force fighter wing equivalents;
- 1-2 Marine Expeditionary Forces;

- 4-5 Navy aircraft carrier battle groups;
- up to 100 bombers;
- Special Operations Forces.[37]

Force Planning into the 21st Century.

Without an agreement on the mission or strategy, force planning in the 21st century will continue to disappoint. Unfortunately, at the present juncture there is little agreement concerning the mission of the armed forces. The on-going debate has two dimensions: shaping and peacekeeping versus warfighting; and current versus future focus. Numerous politicians, defense analysts, and several senior military leaders have concluded that the two MTW requirement should be adjusted to specifically include force-sizing for peace operations. This argument is based on the experiences of the first decade of the post-Cold War period. During that time the operational commitment of U.S. military forces has increased 300 percent, and the vast majority of those deployments have been at the low end of the spectrum of conflict—shaping activities and smaller-scale contingencies, not MTWs. Jeffrey Record argues that the 2 MTW force has little relevance in a world in which a "modern-day version of imperial policing is likely to consume much of U.S. military effort."[38] The most recent articulation of this position was contained in the Phase II report of the U.S. Commission on National Security/21st Century, entitled *Seeking a National Strategy: A Concept for Preserving Security and Promoting Freedom.* The report claims that "the 'two major theater wars' yardstick for sizing U.S. forces is not producing the capabilities needed for the varied and complex contingencies now occurring and likely to increase in the years ahead." It calls for a portion of U.S. force structure to be specifically tailored to humanitarian relief and constabulary missions.[39] Two noted RAND analysts have proposed replacing the two MTW criteria with three simultaneous sizing criteria: force needs for environment shaping; force needs for one tough MTW plus stability operations in other theaters; and force needs for two "moderately difficult" MTWs.[40] Even the Defense Department has begun to waver on the issue slightly. The most recent edition of the DoD Annual Report, in addressing the use of military force in support of primarily humanitarian interests, has removed the previous qualifier that "the U.S. military is generally not the best means of addressing a crisis."[41] This shift in emphasis is further supported by a focus on peacetime military engagement activities as the "best way" of reducing the sources of conflict and shaping the international environment.[42]

Strong voices, however, remain on the other side of the issue. General Shelton, Chairman of the Joint Chiefs of Staff, indicated that the U.S. military should not carve out a portion of its force structure exclusively to handle peacekeeping missions because those operations could quickly escalate into situations that only trained warfighters could handle.[43] Chairman of the House Armed Services Committee, Floyd Spence, in rejecting the Commission on National Security's call to abandon the two MTW yardstick, indicated that he fundamentally disagreed "with those who advocate shifting the composition of our armed forces toward peacekeeping and humanitarian operations at the expense of warfighting capabilities."[44] The need for a versatile and flexible force capable of responding and executing a wide range of missions is clearly recognized. The disagreement concerns where on the operational spectrum should risk be assumed—high end (major theater war) or low end (peacekeeping or humanitarian operations)—or how to posture the force to minimize risk. Force planners will have a hard time developing an acceptable force structure in the absence of consensus on this issue.

Force planners also must resolve the issue of whether to focus their efforts on the current threat or future threats. According to the NDP:

> . . . we must anticipate that future adversaries will learn from the past and confront us in very different ways. Thus we must be willing to change as well or risk having forces ill-suited to protect our security twenty years in the future. The United States needs to launch a transformation strategy now that will enable it to meet a range of security challenges in 2010 to 2020.[45]

Proponents of this view contend that the "revolution in military affairs" (RMA) will have profound effects on the way wars are fought. This model would replace the 2 MTW force with a "silicon-based" superior force that would be smaller and more flexible, emphasizing mobility, speed, and agility. Warfighters would benefit from technological achievements in stealth, precision weapons, surveillance, and dominant battlefield awareness. Most RMA proponents also contend that at present the United States has a threat deficit and therefore can afford to cut force structure and focus on research and development of new "sunrise systems," experimentation and innovation.[46] Critics claim that both the *QDR* and *NDP* failed to propose innovative and long-term changes in the defense program. General Shalikashvili's response to such criticism brings the issue full circle back to risk assessment and how that risk should be allocated over time:

> My admonition was that we need to do what we need to do to remain capable of defending our country and winning our nation's wars. I didn't want to get an award for innovation's sake. I didn't want anyone gambling with our nation's security just so we could be called great innovators.[47]

Since the last *QDR*, the gamble that General Shalikashvili wanted to avoid has only increased in intensity. Most of that tension is associated with the strategy-resources gap mentioned in the introduction to this essay. Eliot Cohen, in a recent *Foreign Affairs* article, emphasizes the funding shortfall:

> For a decade now the Pentagon has not had enough money to replace aging hardware that it uses around the world. Estimates of the shortfall range from $25 billion a year to three

times that figure (the latest assessment from the Congressional Budget Office comes in at $50 billion a year).[48]

The funding shortfall has translated into serious concerns about the readiness of the force to execute the strategy. In recent Congressional testimony, General Shelton indicated that the risk associated with the most demanding scenario has increased. At the same Congressional hearing, Army Chief of Staff General Eric Shinseki stated that "... the increased frequency of mission requirements has had detrimental impacts on the force, especially in terms of operational tempo, personnel tempo, and turbulence."[49] The two MTW focus directly contributes to the undue strains on the force by not necessarily generating the capabilities needed for the full range of missions. Michael O'Hanlon notes the high opportunity cost of over prioritizing the two MTW planning framework,

> . . . keeping a high-priced insurance policy against regional conflict would make it impossible to afford other key defense investments . . . It would also leave us with a force structure not well suited to smaller operations—meaning that ongoing no-fly-zone missions and peace operations will continue to overwork our personnel.[50]

BACK TO THE FUTURE

Force planning has been and always should be a very dynamic process. Consequently, as the strategic environment changes or as the understanding of its uncertainties matures, and as both threat and friendly military capabilities evolve, there should be adjustments to the defense program. It is time to adjust the two MTW force sizing construct. The proposed adjustment that follows does not abandon the requirement for a two MTW capability but articulates that requirement within a more appropriate set of priorities. Because of a more mature understanding of the post-Cold War security environment, the foundations and principles originally associated with the Base Force may be more applicable today than when originally proposed. The logic of the Base Force, combined with the grammar of a force structure sized against a range of missions may provide the "best suited" force planning approach as the United States moves into the 21st century.

A return to the foundations of the Base Force would bring more explicit visibility to the full range of defense requirements, correctly posture U.S. military forces to uphold security commitments to alliance and coalition partners, and better posture forces to respond to the more frequent challenge of SSC operations while preserving the core capability of concurrently deterring multiple MTWs. The Base Force planning construct consisted of four foundations: strategic deterrence and defense, forward presence, crisis response, and reconstitution.[51]

Strategic deterrence and defense focused on the continued relevance of nuclear forces as the ultimate deterrent, and the need for missile defense, at that time referred to as Global Protection Against Limited Strikes (GPALS). Due to the spread of easier access to sophisticated technology over the past decade, the U.S. homeland is more susceptible to attack by hostile entities using a wide range of capabilities from long-range ballistic missiles, to terror weapons, or even information attacks.[52] The expanded range of national security issues that should be considered within this planning category include National Missile Defense, computer network attack and defense, consequence management, and space control. The National Security Strategy recognizes the threat in these areas but discussions about allocating appropriate resources or adjusting force structure are often drowned out by the debate about the readiness and modernization of the two MTW force. Returning to the Base Force construct provides an immediate benefit of giving appropriate recognition to these vital aspects of U.S. security policy.

The other three Base Force foundations concern conventional forces oriented to protecting U.S. global interests. The first of these, forces for *forward presence*, is based on the enduring principle that the presence of U.S. forces in regions vital to U.S. national interests have been key to averting crises and preventing war. In addition, forward-deployed forces are vital to the maintenance of the system of collective defense by which the United States works with friends and allies to protect security interests.[53] The reasons enunciated in support of forward presence remain virtually the same today. The 2000 *Annual Defense Report* highlights the imperative of engagement and the need for cooperative, multilateral approaches to shape the environment and create preferred international conditions. A vital aspect of the military's role in shaping the international security environment is achieved through overseas presence.[54] A brief review of the U.S. regional defense posture in Southwest Asia provides an example of the ends-ways-means of forward presence. The ends are enforcement of United Nations resolutions, ensuring free access to resources, and improving interoperability and regional nations' self-defense capabilities. The ways and means include limited long-term presence forces, a larger number of rotational forces, and prepositioned material and equipment. The 2000 *Annual Defense Report* concludes:

> The close military relationships developed with friends throughout the Middle East and South Asia, complemented by U.S. security assistance programs, contribute to an environment that allows regional states to more readily and effectively support U.S. crisis response deployments. This contribution is integral to U.S. deterrence efforts.[55]

U.S. forward presence has changed very little in the past decade. Forces continue to be deployed in support of vital American and allied interests in Europe, Northeast Asia, and the Pacific, and as mentioned above, in Southwest Asia. What is missing from the two MTW force-sizing construct is specific recognition of that requirement. The Navy represents a qualified exception to this point in that it claims to primarily size its force against forward presence requirements. The Air Force has also recently unveiled its Aerospace Expeditionary Force concept that is designed to facilitate rotational deployments to meet ongoing forward presence requirements. Both services, however, remain strong advocates of their respective "MTW building block" force structures. Concerning naval forces, "Forward presence requirements, and peacetime and crisis response operations . . . are major determinants of naval force needs." Crisis response requirements link back to the two MTW construct so the Navy seems to be hedging its bets on which mission area is a more convincing case for force sizing. Likewise, despite the AEF's goal of reducing unpredictability of forward presence type deployments, the Air Force continues to highlight its capability of deploying "seven to eight fighter-wing equivalents (FWEs) to a distant theater in a matter of days as an initial response in a major theater war."[56] Two times seven or eight comes very close to the total conventional force structure of 19 active and reserve FWEs. For the time being, the services remain wedded to the two MTW force sizing paradigm.

Recall from the earlier discussion of the Base Force in this essay, that forward presence forces were structured and organized geographically—forces for the Atlantic (including Southwest Asia) and the Pacific. Returning to a similar organizing principle would place the appropriate priority on structuring the force to meet the requirements of forward presence and would also recognize unique capabilities needed in different regions. Necessary capabilities should be identified and resourced to respond to the potential threats to U.S. interests in each of these vital regions. Adequate forward presence forces will maintain and strengthen day-to-day alliance and coalition operations, provide the needed capabilities for smaller-scale contingencies in those regions, and constitute an appropriate deterrence posture and enabling capability should the need for a more significant crisis response arise. If by definition the United States maintains forward presence in only those regions of vital importance, then by prioritizing and resourcing these requirements the United

States will be contributing to deterrence and conflict prevention where it counts most.

The third foundation of the Base Force was *crisis response,* which articulated the need for capabilities to respond to regional crises. These capabilities were inherent in the regional forward presence forces as well as the continental U.S. (CONUS) based contingency force. Characteristics of the contingency force included high readiness and the full spectrum of joint capabilities. The size of the force was very similar to the follow-on MRC building block concept. The missions of the contingency force were to complement forward-deployed assets or provide an initial response capability in regions without forward-deployed U.S. forces.[57] In the present context, the contingency force would constitute a significant portion of the nation's major theater war capability. Note, however, that the contingency force would complement the existing warfighting capability of the forward presence forces and thus in combination they would represent the capability to respond to multiple MTW crises. In reality, this posture is little different than that practiced today. Overseas presence forces are counted within the two MTW force requirement. However, by prioritizing the two MTW sizing requirement, the forward presence mission is a lesser-included capability. It seems to be more appropriate to recognize forward presence as the primary mission of these forces and treat their contribution to the second MTW requirement as the lesser-included capability. The apparently prudent logic of the two-war strategy, Michael O'Hanlon argues, "is excessively cautious, in light of the strong U.S. deterrent posture in key regions."[58] Recall that the principal arguments for the two MTW force structure are to credibly deter opportunistic aggressors and fulfill alliance commitments. A recent study on conventional deterrence found that "the most effective military deterrent is the capacity of the defender to repulse an attack and deny the adversary its military objectives at the outset and early stages of an armed confrontation." This study goes on to conclude that effective conventional deterrence "means investing in forward-deployed forces and maintaining a forward military presence."[59]

The final foundation of the Base Force was *reconstitution*. Although the phrase "peer-competitor" was not popularized at the time of the original Base Force, the rationale for a reconstitution capability was targeted to "forestall any potential adversary from competing militarily with the United States."[60] Defense capabilities associated with reconstitution included cadre-type units and defense industrial base initiatives designed to facilitate general mobilization in the face of a reemerged global threat. In addition to the requirements for general mobilization, investments in basic science and high-payoff technologies were also considered reconstitution activities—those defense related activities focusing on generating future capabilities based on new technologies.

Returning to reconstitution as a strategic concept provides two advantages. First, it more appropriately characterizes selected reserve component formations that have limited utility within the two MTW construct. Recognizing these forces as a far-term hedge, or strategic reserve, should clearly reflect a lower priority for resources. The second advantage comes from replicating the current concept of transformation. Transformation activities, such as those associated with developing the Joint Vision 2020 force, should be explicitly recognized as a claimant on force structure. Transformation activities should obviously receive a higher resource priority than other aspects of reconstitution, but it remains appropriate to group them together as approaches to address uncertainties about the future.

CONCLUSION

The post-Cold War era still lacks its own name, but the past decade has contributed greatly to a better understanding of the uncertainties associated with the resultant security environment. The logic of U.S. strategy has remained relatively constant. It recognizes the need for continued U.S. leadership and engagement in the world, coupled with the ability and willingness to respond to threats to U.S. and allied interests, along with appropriate concern about developing future capabilities. Unfortunately, the force-planning grammar associated with this strategy and represented by the two MTW, one-size fits all force planning paradigm no longer captures the full range of force requirements. A return to the foundations and force planning principles of the Base Force provides a better force-sizing framework with which to enter into the 21st century. A comparison of the strategic concepts and force-sizing paradigms associated with the 1997 QDR and a slightly adjusted Base Force 2000 appears below. (Figure 1) Priorities should not be presumed directly from the figure, however it should be clear that Base Force 2000 implies a much broader range of choices. Capabilities to conduct two MTWs remain important but are not necessarily the top priority.

Sizing the Force for the 21st Century

	Strategic Concepts	Force Sizing	
1997 QDR	• Shape • Respond • Prepare	2 MTWs	
Base Force 2000	• Homeland security	Nuclear Forces and NMD Space Control Computer Network Attack and Defense Anti/Counter-terrorism	
	• Forward Presence	Atlantic Pacific Southwest Asia	2 MTWs
	• Crisis Response	Contingency Force	
	• Transformation/ Reconstitution	Science &Technology Selected Reserve formations	

Figure 1.

There are several advantages to adopting the Base Force 2000 concept. First and foremost, it clearly articulates all of the requirements of a multi-faceted defense program which goes far beyond the two MTW requirement. The two MTW construct is too simplistic and thus fails to portray in a convincing manner all defense requirements. The force sizing requirements articulated in the Base Force 2000 construct should allow the defense establishment to both present a more coherent argument for adequate resources while at the same time more efficiently apply available resources to higher priority requirements. Decisionmakers and planners will have a much better framework in which to recognize and make trade-offs between competing defense requirements.[61]

Second, it returns the defense program to more of a capabilities-based force as opposed to the primary threat-based force wedded to two specific MTWs. Recognizing the full spectrum of military capabilities needed to protect U.S. interests, to include providing for homeland security, forward presence, crisis response, and transformation and reconstitution, is the more prudent approach in an uncertain geostrategic environment. The Base Force does not neglect the mantra of being able to "fight and win our nation's wars," nor abandon the two MTW capability. By focusing on an appropriate forward presence posture, however, reinforced by a CONUS-based contingency force, the United States is best positioned to support critical alliances and coalitions and deter or respond to crises that may threaten vital interests. If insufficient resources necessitate a choice between forward presence capability and the second MTW, a higher degree of risk should be accepted in the MTW force. Credibly deterring opportunism is a complementary capability achieved through forward presence and crisis response.

Threat-based planning using the two canonical scenarios of Iraq and Korea suppresses uncertainty and no longer satisfactorily measures the adequacy of U.S. force posture. Furthermore, it is not clear why forces designed for totally different regional scenarios have identical operating (above-the-line) force requirements as expressed in the basic MTW building block.[62] Finally, anchoring the U.S. defense program to two atrophying threats may put the entire program at risk. The Base Force construct allows elements of both threat-based and capabilities-based planning to be applied, broadens the set of planning cases, and emphasizes combinations of capabilities different from those optimized against the two MTW force-sizing construct.[63]

This chapter has presented an argument for a new force-sizing construct but has not attempted to propose a specific force structure. It should be clear, however, that placing a higher priority on forward presence capabilities does not necessarily portend a smaller force, but it certainly could be a different force. Such an approach should, as a minimum, address the issue of low density/high demand capabilities and thus relieve some of the operations tempo (OPTEMPO) pressure. On the other hand, if resources remain limited, which is the likely scenario, the Base Force sizing framework can facilitate debate and potential adoption of appropriate proposals to adjust the force or otherwise address lower priority missions.[64] The Base Force construct clearly articulates all of the requirements of the U.S. defense program and thus allows for a better understanding of priorities, a more efficient allocation of resources across those priorities, and a more informed debate about where to accept risk in the defense program.

The two MTW force planning paradigm has served the United States well, but, as ironic as it may appear, it is time to return to its immediate predecessor and adopt a modified Base Force construct. The Navy claims to be sized by overseas presence requirements and the Marines tout the fact that they are a one-MTW force. It is time to get all of DOD on the same sheet of music. Force planners and strategists involved in the 2001 QDR must rely on an appropriate mix of threat and capability-based planning and examine a broader set of requirements that will allow the United States to achieve its strategic objectives and provide the U.S. political leadership with the appropriate framework from which to judge "how much is enough."

ENDNOTES - CHAPTER 2

1. Harlan Ullman, *In Irons: U.S. Military Might in the New Century,* Washington, DC: National Defense University, 1995, p. 111, identifies three related "vital" questions for force planners:

> What forces are needed strategically and operationally?; What level of capability and what types of force structure are politically and economically sustainable and justifiable . . .?; and How do we safely, sensibly, and affordably get from today's force structure and capability to that of tomorrow and properly balance the threat strategy, force structure, budget, and infrastructure relationships?

2. Certain proponents of the Revolution in Military Affairs argue that the current force is too large and should be cut to afford a greater emphasis on experimentation and modernization. See Andrew F. Krepinvich, Jr., "Keeping Pace with the Military-Technological Revolution," *Science and Technology,* Summer 1994, pp. 23-29; and James R. Blaker, "The American RMA Force: An Alternative to the QDR," *Strategic Review*, Vol. 25, Summer 1997, pp. 21-30. For an earlier argument on the unaffordability of the force, see Don M. Snider, "The Coming Defense Train Wreck . . .," *The Washington Quarterly*, Vol. 19, No. 1, Winter 1996. Michael O'Hanlon, in his book *How to Be a Cheap Hawk: The 1999 and 2000 Defense Budgets*, Washington, DC: Brookings Institution Press, 1998, notes an annual shortfall in the defense budget of between $10 and $20 billion in the next decade. Most recently, Daniel Goure and Jeffrey Ranney argue that the defense department faces an annual shortfall of $100 billion; in *Averting the Defense Train Wreck in the New Millennium*, Washington, DC: The CSIS Press, 1999. Finally, see Harry G. Summers, Jr., *The New World Strategy: A Military Policy for America's Future*, New York: Touchstone Books, 1995, for the argument that the force is totally inadequate.

3. Michele A. Flournoy, *Report of the National Defense University Quadrennial Defense Review 2001 Working Group,* Washington, DC: Institute for National Security Studies, National Defense University, November 2000, p. 7.

4. The Rand Corporation happens to be the principal depository for detailed exposition on force planning methodologies. Among the most

recent works on this subject, refer to the following: James A. Winnefeld, *The Post-Cold War Force-Sizing Debate: Paradigms, Metaphors, and Disconnects*, Santa Monica, CA: Rand, 1992; Richard L. Kugler, *U.S. Military Strategy and Force Posture for the 21st Century: Capabilities and Requirements*, Santa Monica, CA: Rand, 1994; and Paul K. Davis, ed., *New Challenges for Defense Planning: Rethinking How Much is Enough*, Santa Monica, CA: Rand, 1994. In this last work, refer particularly to "Part Two: Principles for Defense Planning," pp. 15-132. Finally, Zalmay M. Khalilzad and David A. Ochmanek, eds., *Strategy and Defense Planning for the 21st Century*, Santa Monica, CA: Rand, 1997.

5. Paul K. Davis, "Institutionalizing Planning for Adaptiveness," *New Challenges for Defense Planning,* p. 81.

6. Quoted in William Kaufmann and John Steinbruner, *Prospects for a New World Order: Decisions for Defense*, Washington, DC: The Brookings Institution Press, 1991, p. 45.

7. Kugler, *U.S. Military Strategy and Force Posture,* p. 35. General Powell defined those missions very broadly:

> We no longer have the luxury of having a threat to plan for. What we plan for is that we're a superpower. We are the major player on the world stage with responsibilities [and] interests around the world.

Quoted in Kaufmann, *Decisions for Defense*, p. 47.

8. Report of the Secretary of Defense to the President and the Congress, Washington, DC: U.S. Goverment Printing Office, January 1991, p. 4; and *National Military Strategy of the United States,* Washington, DC: U.S. Goverment Printing Office, January 1992, pp. 19-24, hereafter referred to as the *NMS 92*.

9. Les Aspin, *National Security in the 1990s: Defining a New Basis for U.S. Military Forces,* before the Atlantic Council of the United States, January 6, 1992, pp. 5-6.

10. Les Aspin, *An Approach to Sizing American Conventional Forces For the Post-Soviet Era,* February 25, 1992.

11. Colin L. Powell, "U.S. Forces: Challenges Ahead," *Foreign Affairs,* Winter 1992/93, Vol. 71, No. 5, p. 41. See also Colin L. Powell, *My American Journey,* New York: Ballentine Books, 1995, p. 438.

12. Dick Cheney, *Annual Report to the President and the Congress*, Washington, DC: U.S. Government Printing Office, February 1992, p. 8.

13. *NMS 92*, p. 11.

14. "The Army Base Force - Not a Smaller Cold War Army," Discussion Paper from the Department of the Army's War Plans Division, dated February 1992. See also Kaufmann and Steinbruner, *Decisions for Defense*, p. 27. The authors make the following point:

> How many contingencies might occur simultaneously, and in how many separate theaters the United States should be prepared to become engaged at any one time, was not made clear. However, the assumption appears to be that the Pentagon should have the capability to deal with at least two major regional contingencies . . .

15. Powell, *My American Journey*, p. 564. Although the supporting analysis behind the Base Force included wargaming the two "canonical-scenarios"—MRC-East and MRC-West, the 1992 National Military Strategy presented the force, as discussed above, as a capabilities-based force. The closest the NMS comes to recognizing a 2 MTW requirement is the following from the Crisis Response section:

> Our strategy also recognizes that when the United States is responding to one substantial regional crisis, potential aggressors in other areas may be tempted to take advantage of our preoccupation. Thus, we can not reduce forces to a level which would leave us or our allies vulnerable elsewhere.

NMS 1992, p. 7. Using the same rationale, 1 year later with the publication of the *BUR*, the two MTW requirement was officially unveiled.

16. Les Aspin, *Report of the Bottom Up Review*, Washington, DC: U.S. Government Printing Office, October 1993. (Hereafter referred to as the *BUR*.) *BUR*: methodology - p. 4; missions - p. 13; force sizing - p. 7. No other administration has provided the degree of transparency in its force planning deliberations as represented by the *BUR*. The detailed wargaming analysis done by J8 is not presented for obvious reasons in an unclassified publication. Nonetheless, contrast this with the history of the Base Force, found in Lorna S. Jaffe, *The Development of the Base Force 1989-1992*, Washington, DC: Joint History Office, Office of the Chairman of the Joint Chiefs of Staff, July 1993, which was not published until at least 2 years after the fact.

17. Richard Kugler, *Toward a Dangerous World*, Santa Monica, CA: Rand, 1995, pp. 212-213. According to General Powell: "It took us nine months to finish the *BUR*, and we ended up again with a defense based on the need to fight two regional wars, the Bush strategy, but with Clinton campaign cuts." *American Journey*, p. 564.

18. George C. Wilson, *This War Really Matters: Inside the Fight for Defense Dollars*, Washington, DC: CQ Press, 2000, p. 15.

19. *Report of the Quadrennial Defense Review*, Washington, DC: Department of Defense, May 1997, p. v. (Hereafter referred to as the *QDR*.) Although sticking with the same basic force requirement, the *QDR* did provide a small degree of innovation in redefining this requirement from 2 MRCs to 2 MTWs. The MRC concept, as first defined in the Bush administration, referred to major regional contingencies. The *BUR* adjusted the term to major regional conflicts, obviously retaining the same acronym. The *QDR* accepted the requirement but changed the name to major theater war (MTW).

20. *Ibid.*, p. 23.

21. *Ibid.*, p. 12.

22. Elaine M. Grossman, "Defense Officials Eye Small-Scale Ops as Organizing Yardstick," *Inside the Pentagon*, March 30, 2000, p. 2.

23. *QDR*, p. 24.

24. *Transforming Defense: National Security in the 21st Century*, Report of the National Defense Panel, December 1997, p. 23.

25. *Ibid.*, p. 1.

26. William Perry, *Annual Report to the President and the Congress*, Washington, DC: U.S. Government Printing Office, 1996, pp. vii-viii. (Hereafter referred to as *DOD Annual Report 96*.) All of the arguments supporting a 2 MTW force posture were reiterated in the *QDR*, pp. 12-13.

27. *A National Security Strategy for a New Century*, Washington, DC: U.S. Government Printing Office, December 1999, p. 19. (Hereafter referred to as *NSS 99*.)

28. *QDR*, p. 13; and *NSS 99*, p. 18.

29. *Ibid.*, p. 12. This point is also made in *NSS 99*, p. 19. The *2000 DOD Annual Report* specifically addresses the need to prevent the

coercion of allied or friendly governments. William S. Cohen, *Annual Report to the President and the Congress,* Washington, DC: U.S. Government Printing Office, 2000, p. 7. (Hereafter referred to as *DOD Annual Report 2000.*)

30. Winnefeld, p. 18. The efficacy of concurrency as a force planning criteria was clearly demonstrated in 1994 when the United States deployed military forces to South Korea and Kuwait in response to threats against U.S. interests. See John F. Troxell, *Force Planning in an Era of Uncertainty: Two MRCs as a Force Sizing Framework*, Carlisle, PA: Strategic Studies Institute, September 15, 1997, p. 22.

31. *QDR*, p. 12. This point is also made in both the *NSS 99*, p. 22, where it refers to the "two theater" force, and in the *DOD Annual Report 2000,* pp. 2-3. The 2000 Annual Report also recognizes the need to guard against wild card scenarios by maintaining "military capabilities with sufficient flexibility to deal with such unexpected events," p. 3.

32. General John M. Shalikashvili, *CJCS Written Statement to Congress*, March 1996, p. 18.

33. *QDR*, p. 12.

34. The "use of plausible, illustrative scenarios against postulated threat forces enables comparisons and analyses to determine the relative values of different forces and capabilities across a range of circumstances." *The National Military Strategy of the United States,* Washington, DC: U.S. Government Printing Office, 1995, p. 17. Refer to the *BUR*, pp. 13-15, for a good discussion of the planning scenarios used to develop the *BUR* force. This discussion addresses most of the common criticisms raised against using these particular scenarios. Robert Haffa, Jr., "A New Look at the Bottom-Up Review: Planning U.S. General Purpose Forces for a New Century," *Strategic Review*, Vol. 24, No. 1, Winter 1996, p. 24, argues that the "work contained in the *BUR* was very much an extension of the scenario-driven methodology that, for the most part, has guided the planning of U.S. conventional forces since the 1960s. . . ." He goes on to state that the "most dangerous near-term contingencies—Korea and the Gulf . . . are among the more stressful to plan against." The *1997 NMS* specifically cites the challenges faced in the Arabian Gulf region and in Northeast Asia. It goes on to say that "even should these challenges diminish this capability [2 MTW] is critical to maintaining our global leadership role." *The National Military Strategy of the United States,* Washington, DC: U.S. Government Printing Office, 1997, p. 15.

35. U.S. Government Accounting Office, *Bottom-Up Review: Analysis of Key DOD Assumptions*, January 1995, pp. 61-64, argued that the scenarios differed from current warplanning assumptions of the CINCs. The DOD response pointed out the differences between illustrative planning/programming scenarios and CINC warplans. The major difference is that the commanders in chiefs (CINCs) are concerned with the present, while the illustrative planning scenarios focus on the future.

36. *DOD Annual Report 96*, p. 5. For a more in depth discussion of the operational phases, refer to *BUR*, pp. 15-17. The *QDR* places particular attention on the need to halt an enemy invasion rapidly. *QDR*, p. 13.

37. It is worth noting that the MTW building block has remained constant over the past several years, with the exception of Army divisions. The original building block specified 4-5 Army divisions. That has since been adjusted to 5 Army divisions. The MTW building block receives mixed coverage in the *QDR*. On the one hand the report claims that "the forces and capabilities required to uphold this two-theater element of the strategy will differ from the Major Regional Conflict building blocks developed in the 1993 Bottom-Up Review." Later, however, concerning the requirements for major theater war, the report concludes that a "force of the size and structure close to the current force was necessary." The *QDR* goes on to claim that a larger force is needed in response to enemy use of chemical weapons or shorter warning times. Finally, the recommended force structure is virtually the same as the *BUR* force. *QDR,* Section III, p. 9; Section IV, p. 7; and Section V, p. 3.

38. Jeffrey Record, *The Creeping Irrelevance of U.S. Force Planning*, Carlisle, PA: Strategic Studies Institute, May 19, 1998, p. 1.

39. The United States Commission on National Security/21st Century, *Seeking a National Strategy: A Concept for Preserving Security and Promoting Freedom*, April 15, 2000, pp. 14-15.

40. Paul K. Davis and Richard L. Kugler, "New Principles for Force Sizing," in *Strategy and Defense Planning for the 21st Century*, Zalmay M. Khalilzad and David A. Ochmanek, eds., Santa Monica, CA: Rand, 1997, pp. 103-104.

41. William S. Cohen, *Annual Report to the President and the Congress*, Washington, DC: U.S. Government Printing Office, 1999, p. 4; and *Annual Report to the President and the Congress*, Washington, DC: U.S. Government Printing Office, 2000, p. 4. The 1999 version contains

the qualifier quoted above as well as other potential constraints on the use of force in support of humanitarian operations. The 2000 version has removed most, but not all of those constraints.

42. General Henry H. Shelton, *Posture Statement before the 106th Congress, House Armed Services Committee*, February 8, 2000. Available from *http://www.dtic.mil/jcs/core/Posture00.html*, Internet.

43. "Shelton Rejects Idea of Separate Peacekeeping Force," *European Stars and Stripes*, April 27, 2000, p. 8.

44. Floyd D. Spence, "Statement of Chairman Spence on the Release of the Commission on National Security/21st Century Phase II Report," *Press Release*, April 19, 2000.

45. *NDP*, p. i.

46. See Andrew F. Krepinevich, Jr., "Keeping Pace with the Military-Technological Revolution," *Science and Technology*, Summer 1994, pp. 23-29.

47. Quoted in Wilson, p. 72.

48. Eliot A. Cohen, "Defending America in the Twenty-first Century," *Foreign Affairs*, Vol. 79, No. 6, November/December 2000, p. 43. The NDU QDR 2001 Working Group refers to a $30-50 billion per year mismatch, Flournoy, p. 7. See also U.S. Congressional Budget Office, *Budgeting for Defense: Maintaining Today's Forces*, Washington, DC: U.S. Government Printing Office, September 2000.

49. Quoted in Anthony H. Cordesman, *Trends in US Defense Spending: The Size of Funding, Procurement, and Readiness Problems*, Washington, DC: Center for Strategic and International Studies, October 9, 2000, pp. 12-14.

50. Michael E. O'Hanlon, "Rethinking Two War Strategies," *Joint Forces Quarterly*, Washington, DC: National Defense University, Spring 2000, p. 12. See also, Flournoy, p. 16.

51. *NMS 92*, p. 6.

52. *NSS 99*, p. 16.

53. *NMS 92*, pg. 7.

54. *DoD Annual Report 2000,* pp. 3, 5.

55. *Ibid.*, p. 12.

56. *Ibid.*, pp. 42, 46.

57. *NMS 92*, pp. 23-24.

58. O'Hanlon, *How to be a Cheap Hawk*, pp. 52-53.

59. Edward Rhodes, "Conventional Deterrence," *Comparative Strategy*, Vol. 19, No. 3, 2000, pp. 246, 250.

60. *NMS 92*, p. 7.

61. The NDU 2001 QDR Working Group emphasized this challenge:

> . . . the QDR presents two broad strategy development challenges. The first challenge is making a strategy-based case for the resources required to meet national objectives at an acceptable level of risk; the second is determining the best strategy possible, given the resources ultimately made available, while explicitly assessing the risks if available resources fall short of the ideal.

Flournoy, p. 18.

62. Don M. Snider, "The Coming Defense Train Wreck . . .," *The Washington Quarterly*, Vol. 19, No. 1, Winter 1996, p. 94. See also Richard Haffa, "Planning U.S. Forces to Fight Two Wars: Right Number, Wrong Forces," *Strategic Review*, Winter 1999, Vol. 27, No. 1, pp. 15-21.

63. Flournoy, p. 11. This NDU report goes on to recommend that:

> the scenario set used for force planning be broadened to include a wider variety of potential warfighting cases, endstate objectives, operational constraints, and joint concepts of operations to ensure that the U.S. military is prepared for the full range of challenges it may encounter in the future.

64. Two detailed discussions of alternative force structures are found in O'Hanlon, *How to be a Cheap Hawk*, Chapter 3, "A New Strategy and Force Structure"; and David A. Ochmanek, *et. al.*, *To Find, and Not to Yield*, Santa Monica, CA: Rand, 1998. The NDU 2001 QDR Working Group discussed the concept of trade space-candidate

approaches to reduce costs while maintaining an acceptable level of risk. Flournoy, p. 8.

CHAPTER 3

REPLACING THE 2 MTW STANDARD: CAN A BETTER APPROACH BE FOUND?

Richard L. Kugler

The standard of sizing U.S. military forces to wage two Major Theater Wars (MTWs) has been a keystone of U.S. defense planning since 1993. Critics are now charging that because this standard allegedly has outlived its usefulness, it should give way to a new approach that is suited to the demands of the coming years. Whether the 2 MTW standard is set aside will not be known until the upcoming Department of Defense (DoD) *Quadrennial Defense Review (QDR)* is complete, and the new administration has formed its defense plans. But one thing already seems clear. Before the 2 MTW standard can be retired, a better replacement must be found, and it has not yet been identified, much less agreed upon.

This chapter offers one such candidate. Doubtless others will appear in the coming months. All new approaches should be evaluated on their merits, in the context of the complex tradeoffs that must be addressed. Clearly U.S. defense planning should not be a prisoner of the past. Just as clearly, change should not be pursued for its own sake, but only when it yields genuine progress. Because the stakes are high, this is a time for deep thinking and careful analysis of the issues and options. Doing so is the best guarantee that regardless of the choice ultimately made, it will be the right one.

My perspective is that of a defense planner who has participated in the analysis of DoD strategic frameworks and force-sizing standards since 1975. Having seen alternative approaches come and go, I grasp the advantages of the 2 MTW standard, but I also am troubled by drawbacks that may become more serious as the new global era unfolds. My main concern is that because world affairs and U.S. national security strategy are changing, the 2 MTW standard may no longer provide a reliable, stand-alone approach to guiding the preparation and use of military forces. Something more comprehensive and inclusive of other high-priority endeavors is needed. The central task, in my estimate, is to craft a new and broader approach of multiple standards that preserves the core strengths of the 2 MTW standard yet also addresses new strategic requirements and priorities. Although DoD has grown comfortable with the 2 MTW standard over the past decade, it cannot afford to stand pat. The time has arrived for it to think and act creatively in this arena. The capacity to innovate at key junctures is the reason why DoD performed well in the past. This remains the case today.

Advantages of the 2 MTW Standard.

The 2 MTW standard is not the first formal approach to force-sizing employed by DoD, and doubtless it will not be the last. One of the toughest challenges facing defense planning is to translate decisions on strategic policy into concrete guidelines for determining exactly how military forces are to be prepared. The role of a force-sizing standard is to help perform this critical task. A force-sizing standard helps determine the size of the U.S. force posture and helps explain the posture's strategic and military rationale in public. In less-visible ways, it also has a major impact on the myriad details of defense planning. It influences judgments about the U.S. defense budget, its subsidiary programs, and its priorities for spending money. It affects how forces are allocated among key missions and commands. It provides a framework for creating plans to deploy and employ U.S. forces in specific contingencies, and for analyzing the capacity of U.S. forces to win the wars they might be called upon to fight. These important functions make DoD's force-sizing standard a matter of considerable significance.

They also greatly complicate the task of choosing a proper standard: never an easy matter in the past and arguably more difficult today because the Cold War's clarity has been replaced by a murky, confusing world.

The 2 MTW standard was adopted in 1993 and reaffirmed in the QDR of 1996 because it was deemed capable of meeting the emerging demands of the post-Cold War era, including deterrence of regional wars that might break out as the old bipolar order unraveled. It permitted the Clinton Administration to trim about 10 percent off the "Base Force" inherited from the Bush era. The Pentagon came away feeling that while the resulting posture offered less margin of safety, it likely would be large enough to get the job done. As the 1990s unfolded, the 2 MTW standard endured as a serviceable doctrine that was neither widely admired nor hotly opposed. The fact that it has survived this long is testimony to its staying power and lingering appeal: attractive features at a time when consensus on defense policy is hard to come by. Today's criticisms of it are being launched mostly by defense specialists, not by Congressmen, other political figures, or protesters in the streets. If it is to be retired, it will not go quietly into the night without a stiff debate. Any replacement for it will need to show not only superior substantive qualities but also a similar capacity to command consensus inside and outside the Pentagon: a hard act to follow.

This standard was created in the aftermath of the Persian Gulf war, but before ethnic warfare had fully exploded in the Balkans. It postulated that U.S. military forces should be large enough to wage two large-size regional wars that might erupt with little warning and unfold in overlapping time frames. The two contingencies most commonly cited were a renewed Iraqi attack on Kuwait and Saudi Arabia, accompanied by a North Korean assault on South Korea. Both were dangerous events that would menace U.S. vital interests, activate U.S. security ties with close allies, and require major U.S. force contributions. The 2 MTW standard calculated that if U.S.

forces are sufficiently large and capable of winning two such wars, they will be able to deal with the biggest existing threats on the world scene and be prepared to handle other wars and crises that might unexpectedly occur.

The 2 MTW standard thus was anchored in big-time warfighting. It did not mean that U.S. forces could be employed only to wage MTWs. During peacetime, it allows for U.S. forces to be used for other crises and contingencies: originally called "Lesser Regional Contingencies" (LRCs) and now dubbed Smaller-Scale Contingencies (SSCs). Yet it also made clear that forces employed for these purposes must remain on a tetherhook, primed to disengage and redeploy if MTW events necessitate their presence. The 2 MTW standard thus was meant to be flexible in determining how U.S. forces might be used, but nonetheless, it was firm about the main purpose of the force posture: being prepared for MTWs and engaging in other operations only on a "by exception" basis. In using MTW requirements to judge military adequacy for the 1990s, it declared that a 2 MTW posture would satisfactorily answer the perennial question: How much is enough?[1]

The 2 MTW standard remains alive today, and continues to enjoy support in many quarters that value its advantages. It provides the officially sanctioned rationale for virtually all U.S. active-duty combat forces and many reserve component units as well. This is the case because these forces would be needed to meet the weighty demands posed by two MTWs erupting nearly at the same time. Basically, the standard postulates that about one-half of U.S. forces are needed for one MTW, and the other half are needed for the second MTW. The current U.S. posture includes the combat forces listed in Table 1. The only forces falling outside the 2 MTW standard are some Army National Guard units, which provide a low-cost, mobilizable hedge against more demanding events. A typical MTW commitment would include about 6.5 Army and Marine divisions, 10 USAF fighter wings, and up to 4-5 carrier battle groups (CVBGs). These combat forces would be moved by sizable strategic mobility forces: air transports and sealift ships. Accompanying the combat forces would be command staffs, sizable logistic support units, and war reserve stocks. Total deployed manpower for a single MTW posture would be about 400,000-450,000 troops from all services. Clearly this is a large and powerful combat force, capable of major defensive and offensive operations.

	Active	**Reserve Component**
Army Divisions (Separate Brigades)	10 (2)	8 (18)
Marine Divisions & Air Wings	3	1
Air Force Fighter Wings	12	8
• Bombers	163	27
Navy Battle Force Ships	301	15
• Carriers/ARGS	12/12	–
DoD Military Manpower	1.35 million	865,000

Table 1. U.S. Defense Posture (2000).

The advantages of the 2 MTW standard are severalfold. It serves as a clear reminder that big wars similar to Desert Storm can still occur because even though the United States no longer faces a global threat akin to the Cold War, it confronts strong regional enemies potentially willing to commit aggression. This standard also signals the continuing U.S. willingness to defend its vital interests and protect its close allies. Its emphasis on warfighting creates compelling reasons for DoD to preserve the world's highest-quality military forces, with high readiness, modernization, and sustainment. The two wars contemplated by this standard require good mobility forces, joint operations, and the mix of ground, air, naval forces contained in the U.S. posture. The 2 MTW standard also provides support for the Revolution in Military Affairs (RMA), JV2010/2020, and DoD's plan to begin a new era of procurement in a few years.

In addition to usefully linking U.S. forces to clear threats and plausible wars, the 2 MTW standard reduces calculations of force requirements to a simple numerical algorithm. By proclaiming the need for a two-war posture, not one war or three wars, it has boiled defense planning down to a single-point solution. It has helped build a broad political consensus for the current posture, establishing a ceiling over the posture and a floor under it. The 2 MTW standard achieves this end with arithmetic proclaiming that more forces would be superfluous and fewer forces would be inadequate. Seasoned military officers and operations research analysts may recognize that reality is more complex than this formula. But the 2 MTW standard thus far has gotten the job done in the public arena in ways that have helped insulate the Pentagon from Washington's political struggles. Meanwhile, it has allowed DoD to resolve its internal debates by focusing on two clearly defined wars, whose postulated features have been developed in satisfying detail to permit analysis of plans, programs, and budgets.

Most important, the 2 MTW standard has bequeathed a force posture that thus far has adequately met U.S. security requirements. While U.S. forces often have seemed over-stretched when mounting peacekeeping missions and related operations in recent years, no enemy has seen opportunity to launch MTW-style aggression in the Persian Gulf or Northeast Asia. Perhaps this situation owes to peaceful international conditions. But to an important degree, it may also owe to the fact that the U.S. military, despite its problems and shortfalls, genuinely possesses the capacity to work with allied forces to inflict decisive defeat not just on one enemy, but on two enemies at once. To be sure, this capacity is not perfect. But even so, it is sufficiently strong to ensure eventual victory in both cases and crushing defeat for enemies. Because U.S. wartime operations in one theater do not open the door to unopposed aggression in another theater, the 2 MTW standard underscores deterrence to a significantly greater degree than could be achieved by a one-war standard. Meanwhile, U.S. military forces generally have been available for other lesser contingencies, including the Kosovo conflict, that have occurred in recent years. While the Pentagon sometimes has grumbled about such events, no major crisis requirement has gone unmet. For these reasons, many defense analysts agree that a 2 MTW force posture is a cost-effective choice: it provides adequate military preparedness and safety, while not draining the federal treasury dry.

Criticisms of the 2 MTW Standard .

Despite its advantages, the 2 MTW standard has been bombarded by mounting criticisms in recent months and by fault-finding on multiple grounds. Supporters of the 2 MTW standard assert that it has been misinterpreted and taken out of context. Perhaps so, but these misinterpretations have acquired a life of their own, to the point where this standard now means something that may not originally have been intended. Some critics have been calling for larger or smaller forces, but the main complaints have focused not on force size, but instead on the standard's strategic rationale and priorities for preparing and using forces. The following portrayal focuses on eight critical arguments being advanced, not the personalities making them. In this chapter's view, no force-sizing standard can be perfect, but most of these criticisms are valid to one degree or another. The looming question is: What should be made of them? Are they mere chinks in the 2 MTW standard's armor, or are they strong enough to justify overturning it? This question is for readers and the U.S. Government to answer.

One criticism is that the 2 MTW standard allegedly fails to focus on the normal business of the Department of Defense and U.S. military forces. It is totally preoccupied with preparing U.S. forces to fight two major regional wars at the same time. To be sure, staying prepared for such calamitous events is critically important. It provides DoD a compelling rationale to keep U.S. forces ready for prime time and to improve them as opportunities arise. But regional wars—even one war at a single time, to say nothing of two wars—occur infrequently. The vast majority of the time, U.S. forces are engaged in far-flung, demanding activities of a different sort: e.g., training, developing collaborative practices with allies, reaching out to new partners, patrolling troubled areas, peacekeeping, striking at terrorists, and conducting minor crisis interventions. The 2 MTW standard implicitly assumes that if U.S. forces are prepared to wage regional wars, they will be able to perform all these other missions. But is this truly the case? Because critics fret that it may not be true, they worry that an exclusive focus on MTWs can result in other vital force needs and program priorities being neglected.

A second, related criticism is that because the 2 MTW standard is preoccupied with fighting wars, it allegedly says nothing about the larger role played by U.S. forces in carrying out the national security agenda abroad. To be sure, the U.S. national security strategy of shaping, responding, and preparing calls attention to the important peacetime purposes of U.S. military forces. But the connection of these strategy precepts to the 2 MTW standard seems tenuous. This especially is the case because the quite-important shaping function plays no role in the analytical process by which DoD gauges its force requirements for the 2 MTW standard. The wrong-headed implication, critics allege, is that U.S. forces exist mainly to wage war and only distantly to help achieve key political and strategic goals in peacetime.

This self-preoccupied focus on warfighting allegedly leaves DoD force planning curiously disengaged from the central thrust of U.S. foreign policy in today's world: keeping the peace and molding the future international system. Several observers, including the Secretary of State, have expressed mystification at this detached attitude, which was not the case during the Cold War, when the Pentagon had a good reputation for grasping the political purposes of military power. The ultimate effect, some critics fear, will be to weaken the U.S. military's credibility and importance in the public eye. They argue that the Congress and American people may not be willing to continue supporting a military capable of waging two hypothetical wars with the most vigorous war plans and highest-technology assets imaginable. But most fair-minded observers will be more-inclined to support a force posture of this size and strength if it powerfully contributes to keeping the peace and influencing how countries respect U.S. interests. Unfortunately the 2 MTW standard does not call attention to this important role of U.S. military power.

Beyond this, the new administration will need to ask an all-important strategic question: How many U.S. forces are needed to keep Europe stable, to dampen the Middle East's chaotic affairs, and to guide Asia toward a new security architecture as China's power grows? The 2 MTW standard cannot address this question, much less answer it. Its allegedly lame response is that forces sized to fight two regional wars presumably will be big enough, and properly configured, to achieve these core political goals. If this proves to be the case, it will be by accident, not design. What if U.S. force needs for these goals prove to be greater than, or different from, requirements for warfighting? In this event, the 2 MTW standard could leave U.S. national security strategy flying blind or at least weakened. During the Cold War, U.S. forces were sized and publicly justified not merely to wage war, but also to contribute to such larger strategic precepts as containment, deterrence, forward defense, flexible response, and alliance preparedness. Critics allege that the time-honored value of this dual focus on peacetime missions and warfighting has been lost in the 2 MTW standard.

A third criticism of the 2 MTW standard is that it anchors the U.S. defense rationale too single-mindedly in fleeting threats and too narrowly in an outdated form of threat-based planning. Iraq and North Korea have served as the principal threats of the past decade. But they are not necessarily permanent fixtures for the coming decade. Indeed, they could fade from the scene quickly if Saddam Hussein's government is overthrown and if the Korean peninsula unifies. What would happen to DoD's force requirements then? Would a major disarmament be possible? Or will new threats appear on the scene? Indeed, should the United States remain heavily armed even in a setting where no major wars loom on the immediate horizon because new dangers could eventually appear, far faster than a disarmed United States could prepare for them?

More fundamentally, this criticism alleges that the 2 MTW standard is trying to use two middling-sized adversaries, with backward economies and no alliance ties to each other, to create a replacement for the role of permanent, multi-theater enemy played by the Soviet Union during the Cold War. To be sure, permanent enemies help make things simple, unchanging, and free of controversy over such thorny issues as goals and priorities. They allow defense planning to be reactive, sparing it the need to be proactive. The problem, critics allege, is that this type of threat-based planning has gone the way of the Cold War. In today's fluid world, the primary task is to advance U.S. interests, not to ward off ceaseless and growing military threats. Permanent big-time enemies no longer exist. Whereas today's enemies may be unthreatening tomorrow, countries that seem quiescent today may become major trouble-makers a few years from now. The task facing U.S. defense planning allegedly is to remain flexible in ways that deal with this ever-shifting strategic scene, rather than rigidly lock itself into a threat-based framework that may bear little relationship to how the future actually evolves.

Most critics likely would agree that threat-based planning should remain a contributing factor in U.S. defense planning. But they also argue that it should not be carried to the point of obscuring the need for the United States to stay well-armed in a still-dangerous world irrespective of the comings and goings of particular threats. The United States is a global power with far-flung interests, security commitments, and involvements in troubled regions that are capable of unexpectedly exploding into conflict almost overnight. The two major wars of the past decade, in the Persian Gulf and Kosovo, both caught the United States by surprise: neither Iraq nor Serbia were identified as full-fledged threats before they suddenly invaded their neighbors. Given these realities, the current U.S. defense posture of 13 active Army and Marine divisions, 20 USAF fighter wings, and 12 Navy CVBGs and ARGs is a sensible capability even in a setting where no war-producing threats immediately exist. Over-reliance on threat-based planning risks losing sight of this key strategic judgment, thereby exposing the U.S. posture to damaging turbulence in a fluid setting where threats appear and disappear with regular frequency. U.S. defense planning requires stable continuity: in today's world, the old form of threat-based planning allegedly fails to provide it.

A fourth criticism is that the 2 MTW standard is too-beholden to the allegedly faulty premise that two regional wars menacing U.S. interests may occur simultaneously. While the world remains a dangerous place, the reality is that for the past 50 years, regional wars involving U.S. forces have occurred only one at a time, not two at once. The same pattern, this criticism alleges, is likely to hold true in the future. U.S. forces won the Persian Gulf war and the Kosovo war so decisively that politically isolated countries like Iraq and North Korea seem unlikely to be willing to risk major war with the United States and its allies. Even if one country or the other is tempted to commit aggression, they are unlikely to launch concurrent attacks because they are not allies, are located far apart, and respond to dissimilar situations. When the time is ripe for one country to attack, it likely will not be ripe for the other. Seasoned defense planners grasp the need to be prepared for two wars so that a strong response can be mounted in one region without fear that an enemy in another region will be given opportunity to attack. But to the man in the street, the expensive act of spending many billions of dollars on being ready for two wars can come across as a wasteful exercise in worst-case planning, if not bureaucratic self-indulgence. Possessing large forces as a back-up insurance policy against two wars makes sense provided these forces are also needed for other high-priority purposes. But possessing them solely or even primarily for such insurance is something else again.

A fifth criticism is that the 2 MTW standard allegedly creates blinders to theaters, missions, and contingencies that lie outside its focus on regional wars in the Persian Gulf and Northeast Asia. The strategic reality is that although this standard focuses on only two theaters, the United States has major security involvements and military commitments in three theaters, and occasionally carries out operations in others as well. When the 2 MTW standard was initially adopted, it seemed to relegate Europe and NATO—previously a centerpiece of U.S. defense strategy—to the backwaters. Even so, the United States chose to keep fully 100,000 or more troops in Europe (its biggest overseas troop deployment) and to continue accepting major roles in NATO's war plans for projecting large forces in and around Europe. But the forces allocated to these war plans, including troops in Europe and reinforcements from the United States, were also assigned major missions in the defense of other theaters. To put matters mildly, this perplexing behavior left Europeans confused. Their confusion grew when the United States subsequently announced policies to enlarge NATO and to adapt alliance forces to new power-projection missions in which European units would be expected to carry out modern doctrine with high-technology U.S. forces. One of the great ironies of the late 1990s is that when a major war erupted in Kosovo, it occurred in the one theater apparently assumed by U.S. defense planning to be immune from the threat of war. The United States responded effectively by concentrating large air and naval forces for NATO's campaign, but the step required controversial decisions to deploy some units that were assigned to other regional commanders and war plans.

The problem of allegedly inflexible defense plans that can leave regional commanders in chief (CINCs) high and dry goes beyond Europe. A few years ago, the United States was required to deploy aircraft carriers toward Taiwan in order to help dampen an impending crisis between China and Taiwan. The carriers were sent, but they had to be temporarily extracted from forces earmarked for the defense of South Korea. In the coming years, CINC-Pacific (CINCPAC) may face regular needs to deploy forces to multiple spots in Asia and the Pacific for peacetime shaping missions and crisis response. If force commitments to the Korean contingency are treated inflexibly, in ways that hamstring not only forces stationed in South Korea but also units based elsewhere in the Pacific and the United States, CINCPAC's ability to respond could be impeded. Indeed, all three major regional commanders—Commander in Chief-Europe (CINCEUR), CINCPAC, and Commander in Chief-Central (CINCCENT)—could face constraints on their ability to carry out missions and respond to crises that depart from the canonical scenarios of the 2 MTW standard.

For all key regions, one concern is that future big wars may not take the shape of today's MTW contingencies, which contemplate major ground attacks across the borders of U.S. allies. Because proliferation is accelerating, such wars may involve use of weapons of mass destruction (WMD) systems and asymmetric strategies aimed at clouding the political situation and slipping the U.S. military punch. Even some wars employing conventional forces in traditional ways may mostly take the form of long-range air and naval operations, and they may not be waged over control of borders and territory. The implication is that while the United States should remain prepared for big wars, its defense planning should examine the full spectrum of possible conflicts, not merely those that replicate Desert Storm and occur only in the Persian Gulf and Northeast Asia. A force posture optimized to wage today's two MTW conflicts does not guarantee success if the big war that actually erupts is so different from these conflicts that U.S. forces are unable to meet its demands. For example, current U.S. defense plans call for swift deployment of small forces capable of sending a deterrent signal and contesting enemy advances, followed by buildup of much larger forces, over a period of several weeks, for an eventual counterattack. This response worked in the Persian Gulf war, but what if a future war requires deployment of medium-sized strike packages, with counter-WMD assets, at a faster pace than now planned? Worrisome events of this sort are reason for thinking beyond the narrow confines of today's MTW contingencies.

An equal concern arises over the ability and freedom of U.S. forces to react to wars and operations that are significantly smaller than today's MTW contingencies. The controversies that have erupted in response to the U.S. military performing peacekeeping missions make clear that the problem of constraints created by MTW commitments remains real, not merely theoretical. On the surface, force commitments for such missions seem modest: normally, no more than 2-3 percent of active U.S. military manpower. Yet such missions disrupt some training for warfighting, cause other forms of turbulence, and often result in the sustained deployment of units that have assignments in MTW war plans. Some observers complain about U.S. forces allegedly being run ragged by non-warfighting missions and express nervousness about the damage done to MTW preparedness, which is central to U.S. defense strategy. Skeptics counter by deriding the prospect of the entire U.S. military being kept of the shelf, standing guard against two regional wars that are unlikely to erupt, and being unavailable to deal with important events and missions that actually are occurring. Thus far, DoD has always responded flexibly by making available the necessary forces for each operation. But to critics, the debates that have accompanied these actions reflect an inherent flaw in U.S. defense planning, one that can result in an inability to see the forest through the trees. After all, they say, military forces are created in order to be useful when they are needed, not treated as a precious asset that can be applied only in a few extreme cases that seldom, if ever, occur.

A sixth criticism is that the force calculations used for the 2 MTW standard seem questionable in ways that can result in the tail wagging the dog. Why is it that the same number of forces—one-half of the U.S. posture in each case—is needed to wage two wars that differ vastly in enemy and allied forces, the terrain, and strategic circumstances? The idea that U.S. forces of 6-7 divisions, 10 USAF fighter wings, and 4-5 carriers are needed to defend the Persian Gulf seems reasonable. But the same judgment applies less clearly to Korea. During the Cold War, when the situation was more precarious than now, the United States planned to defend South Korea with smaller forces. The idea was to rely on the South Koreans to meet the bulk of force requirements, with the United States providing only modest air and naval support. The South Koreans, blessed by a booming economy and a population of 46 million, seem capable of performing this task today. In particular, their army of about 25 divisions is amply large to populate densely a peninsula that is only 200 kilometers wide and marked by rugged terrain and prepared positions. The notion that large U.S. ground forces might be needed to launch a powerful counterattack late in the battle seems a stretch: it assumes that most South Korean units have been destroyed but the North Korean army has suffered only lightly. Surface appearances suggest that a smaller U.S.

force allocation might suffice. This step could free some units for missions other than the 2 MTWs, but it also would mean that a sizing concept broader than 2 MTWs is being used. Allocating this many forces to the Korea contingency maintains the simple clarity of the 2 MTW standard, but perhaps at the cost of the flexibility needed to perform other missions. If the U.S. force posture lacks adequate flexibility owing to the rigidity of its own planning standard, the tail seems to be genuinely wagging the dog.

A seventh criticism is that DoD allegedly does not take being fully prepared for two regional wars seriously in its own programming and budgeting, even though failure to do so can come across as a major deficiency in U.S. defense preparedness. This criticism has its origins in the reality that at a time of fiscal constraints, DoD naturally sets priorities in how it allocates funds. Programs of primary importance normally receive full funding, but less-critical programs sometimes are short-changed to one degree or another. In the eyes of this criticism, DoD seems intent on being highly prepared for one regional war while attending to readiness and modernization, but is willing to accept some shortfalls in being able to mount a second MTW response. Critics point to alleged shortfalls in strategic lift, war reserve stocks, and specialized assets, and in the practice of relying on reserve component forces to provide logistic support. Some critics decry the damage allegedly done to the top priority of U.S. defense strategy: being prepared for two regional wars. Others question the need to be fully prepared for two wars. What unites both camps is that they are fingering an alleged disconnect between strategy and budgets: the kind of disconnect that, if carried too far, can result in serious trouble for U.S. defense preparedness because of failure to establish clear priorities.

An eighth criticism, normally advanced less strongly than the other seven, holds that the 2 MTW standard misjudges the size of the U.S. defense posture that should be maintained in the coming era. Some critics assert that the 2 MTW standard inflates U.S. force requirements. They argue for smaller forces either to trim allegedly wasteful defense spending or to focus additional money on military pay, readiness, and procurement. Other critics allege that the 2 MTW standard underestimates force needs. They call for a bigger posture in order to carry out missions that are deemed likely to grow in the coming years. Their primary goal is to assemble enough additional forces to safeguard against MTWs while carrying out other missions, including peacekeeping and peacetime shaping. Critics from this school thus agree that the 2 MTW falls short in its core function of accurately gauging force requirements, but they disagree sharply on the remedial steps to be pursued.

How seriously should these eight criticisms of the 2 MTW standard be taken? To those clamoring that the 2 MTW standard should be junked, they are a devastating indictment. To those who still see major advantages in the 2 MTW standard, they are not sufficient to justify throwing out the baby with the bathwater. But even supporters of the 2 MTW standard would be hard-pressed to deny that its problems should be fixed, if possible. The real issue is not whether the 2 MTW standard is fatally flawed or still serviceable, but whether a better replacement can be found. This is an issue that requires careful thought, for while criticizing the 2 MTW standard is easy, designing something better is harder.

Toward a New Approach to Force-Sizing: Assessing the Options.

A fair appraisal of the 2 MTW standard is that it made sense in 1993 and for several years thereafter. But in the period since it was adopted nearly a decade ago, the world has changed a great deal, and U.S. foreign policy and national security strategy have also changed. The early 21st century promises even greater changes. In the fluid era ahead, DoD will need to gain maximum strategic mileage from resources that may be less than ideal. A main problem is not that the 2 MTW standard grossly misidentifies total force needs, but that it can give rise to a narrow, rigid focus on a single strategic purpose at a time when a flexible, adaptable focus on multiple strategic purposes will be needed. The emerging situation requires an approach to force-sizing and the other dimensions of defense planning that addresses the future, not the past.

Some of those who favor retaining the 2 MTW standard argue that the United States can solve its defense dilemmas by focusing its military forces solely on being ready for major wars while eschewing other burdensome missions. Presumably responsibility for these missions would be assigned to allies and partners. While this approach sounds appealing at first blush, it breaks down when its adverse consequences are considered. The core problem is that U.S. military forces would not be available to support national interests in many non-war situations that will have a major impact on how the international system evolves. Too often, U.S. forces would be left standing on the sidelines, and U.S. foreign policy would be denied one of its most important instruments. U.S. influence abroad would decline, and sooner or later, U.S. forces might be compelled to fight more regional wars than otherwise will be the case.

Could allies and partners be relied upon to act in ways that protect American interests? Almost certainly not, skeptics allege. These countries have their own interests to serve, their military forces are not adequate for serious power projection missions, and they rely upon U.S. leadership to mobilize their own capacity to act. The idea of exercising restraint in the use of U.S. forces, demanding greater allied contributions, and forging a better division of labor with them makes sense. But withholding U.S. forces from critical new-era missions that are here to stay is an ostrich approach to the coming era, one that seems doomed to fail and backfire.

What can be done to keep U.S. forces properly focused on MTWs while still being available for other missions? If DoD's upcoming *Quadrennial Defense Review* decides to retain the 2 MTW standard, greater flexibility perhaps can be gained by trimming allocations to the Korea contingency and using the resulting forces to form a strategic reserve for other missions. Essentially a new sizing standard would be created: forces for two MTWs and one SSC. The current force posture might be able to carry out this approach in the short term, but over the long haul, larger forces could be needed to make it viable. The reason is that a slimmed-down Korea commitment is a generic MTW posture on the cheap. In any event, this approach amounts to applying a bandage to one key problem, but it does not solve all of the problems arising from the 2 MTW standard. A more fundamental change, aimed at replacing the 2 MTW standard with a new and innovative approach, may be needed.

The act of contemplating how to carry out such a change must begin with a clear-eyed appraisal of how the world is evolving in the early 21st century. In essence, world affairs are being transformed by globalization. The growing tempo of international activity in trade, finances, information, technology, and values is drawing once-distant regions closer together in time and space, making them more interdependent. As a consequence, the United States must learn to think and act globally, for its interests are enlarging, and it is now becoming vulnerable to events at the far corners of the world, including places that were once deemed outside the perimeter of its defense planning.

Global economics and security affairs now share center stage, and they are interacting in complex ways to shape the future. While the ultimate outcome is impossible to know, new opportunities are being created, but so are new dangers that are more serious than is often realized. A positive trend is that the U.S.-led democratic community, which now covers nearly one-half of the world, is benefiting from globalization, becoming wealthier, and drawing closer together in peaceful cooperation. But the tumultuous regions comprising the rest of the world are a different matter. Already beset by troubled politics and economics, many of these regions are being both helped and harmed by globalization. The consequence is growing chaos and unstable security affairs that have the potential to produce heightened military tensions, conflict, and war.

Especially menaced is the so-called "southern belt," stretching from the Balkans in Europe, through the Middle East and Persian Gulf, across South Asia, and along the great Asian crescent from Southeast Asia to Japan. U.S. military forces will need to remain committed in such traditional places as Europe and Northeast Asia, while being capable of operations in Africa and Latin America when the need arises. But the stressful events of the past decade, including two wars, make clear that in the coming years, U.S. forces likely will be called upon to operate along the southern belt more often than during the past. Indeed, the southern belt could become a new main geographical focus of U.S. military activity. If so, this situation will bring about major changes in how the United States thinks about using military power in order to protect its interests, control chaos, and foster stable security affairs so that the progress-producing side of globalization can take hold.

Owing to new geography and other trends, including WMD proliferation, future U.S. military strategy seems destined to be less positional and less continental than in the past. In the coming decade and beyond, it will be animated more by maritime concepts, flexible joint operations, and adaptive responses than before. It will be placing a growing premium on versatile peacetime strategic shaping and on swiftly projecting military power in varying forms to ever-shifting places, some of them well-removed from existing bases and facilities, in crisis and war. It will be confronting new dangers, new threats, new conflicts, and new adversaries intent on frustrating U.S. strategic designs. Meanwhile, it will be presiding over U.S. military forces that themselves are being transformed in response to the information era, the RMA, new doctrine, and coming procurement efforts.

The core issue facing DoD is one of crafting a new approach to defense planning that reacts sensibly to these major changes, accurately measures force needs and priorities, and offers a credible strategic rationale that can endure—both inside the Pentagon and in the public arena. The act of adopting a new approach is one that should be pursued carefully, for many issues must be considered. Broadly speaking, there are three approaches to force sizing: new contingency-based standards, capability-based standards, and strategy-based standards. Contingency-based standards would continue to size and design U.S. forces on the basis of wartime needs: e.g., enough forces for 1.5 MTWs or 2.5 MTWs instead of today's 2.0 MTWs. Capability-based standards would aspire to determine the force characteristics needed for a wide spectrum of operations (e.g., sufficient land forces to provide a robust mixture of infantry, armored, mechanistic, and air assault units). The same applies to air and naval units. Strategy-based standards would look beyond wartime contingencies and combat capabilities to determine the forces needed to carry out the key precepts of national security strategy. All three approaches have their advantages and disadvantages. The tradeoffs need to be evaluated carefully before making a decision. The key point is that today's standard is not frozen in concrete. If another approach is deemed better, the door can be opened to adopting it.

Without pretending to settle this issue, this chapter reasons that strategy-based standards, supplemented by analysis of contingencies and capabilities, may work best. This approach's key advantage is that it would anchor force planning in a stronger strategic foundation than solely being prepared for hypothetical wars. This approach was used successfully during the Cold War, during which U.S. forces were sized to carry out national strategy through a broad spectrum of capabilities, while also responding to the dictates of wartime contingency plans. Back then, U.S. defense strategy was often called "flexible response," a political and military term that said a great deal about the virtues of not becoming too locked into single prepared scripts that overlooked something important or could be overturned at the drop of a hat. After all, Dwight Eisenhower once wisely said that "plans are nothing but planning is everything." In the old but new approach put forth here, U.S. forces would be sized to help carry out the three key precepts of national security strategy: shaping, responding, and preparing—or their successors. Once this key task is accomplished, forces can be fine-tuned to perform specific contingency plans and provide a flexible portfolio of assets.

Illustratively, a strategy-based approach can be brought to life by anchoring U.S. defense plans in a nested hierarchy of three new standards that together provide a reliable measure of enduring military needs and a credible strategic rationale for the resulting posture. The first two standards are primary: chief mechanisms for determining force needs because they focus on the most common strategic missions of U.S. forces and high probability events. The third standard is complementary, ensuring effective forces in more demanding, but less-probable events. The idea behind this strategy-based approach is to determine requirements and priorities for each of these standards, and then to choose the kind of balanced, multifaceted force posture that does the best job of serving all three standards.

Instead of replacing today's single standard with a different single standard, this approach thus creates three sensible standards, and calls upon DoD to anchor force sizing and planning in a synthesis of all three of them. It asks a simple question: Why have one standard when three standards are needed to provide a reliable guide to strategic planning? After all, if modern business corporations can grapple with the task of assessing multiple strategic purposes, so can DoD. Countless books on the theory of corporate decisionmaking have argued that sound strategic plans are best made not by rigidly employing one line of reasoning but by employing several lines and blending them together in sensible ways. Indeed, DoD regularly considered multiple standards during the Cold War and used them to make a steady stream of wise decisions. It needs to recover this lost art. The three new standards are:[2]

Standard 1: Forces for Normal Global Missions. Its purpose is to ensure that during conditions short of major war (i.e., 95 percent of the time), the principal U.S. regional military commands—especially EUCOM, CENTCOM, and PACOM—always have enough forces available to them to perform their normal duties, such as training, working with allies, outreach, peacekeeping, and responding to small-to-medium crises and conflicts. Such forces would include overseas-stationed assets plus units based in the United States that could be drawn upon when necessary. Illustratively, this standard might assign or make readily available a posture of three divisions, five fighter wings, two CVBGs, and an ARG to each of these three commands. Remaining forces would be withheld as a strategic reserve under national command, for flexible use in other missions and regions.

Standard 2: Forces for a Single MTW, while Performing Normal Missions Elsewhere. Its purpose is to ensure that U.S. forces swiftly can concentrate to win a single big regional war in varying places, while not seriously denuding the other major CINCs of forces needed to carry out their normal missions. In event of a Persian Gulf war, for example, this standard would commit forces already assigned to CENTCOM plus draw upon the strategic reserve to create an adequate wartime posture. Meanwhile, EUCOM and PACOM would retain control of most or all of the forces normally assigned to them. Thus their normal operations would not be severely degraded. A similar calculus would apply to wars in other theaters.

Standard 3: Forces for More Wars, or Different Wars, or Bigger Wars. Its purpose is to ensure that in event of more demanding wartime situations than Standard 2, U.S. forces will be adequate to the task if full use is made of the opportunity to concentrate them and employ them adaptively. This standard would examine needs for two MTWs in overlapping time frames. It also would examine force needs should a different war or a bigger war, well-larger than today's MTW conflicts, erupt.

The main effect of these standards is to establish a new, broader frame of reference for articulating and pursuing the main strategic purposes and priorities of force planning. Above all, they place the need to be ready for two MTW conflicts in a larger context and they devote separate treatment to other concerns of equal or greater importance. Essentially, they say that U.S. forces should be made capable of carrying out three strategic purposes: (1) normal peacetime missions and crisis-response duties, which themselves are demanding; (2) fighting one big war, while keeping other theaters stable; (3) in extremis, waging two big wars or similarly demanding conflicts at the same time. Rather than assuming that Standard 3 preparations will produce adequate forces for Standards 1 and 2, they call for a careful examination of force needs and program priorities for all three standards on an individual basis.

These standards also provide a more diverse and potentially better way to think about how U.S. military forces are combined together and used in the shifting array of operational circumstances likely to confront them. The current standard provides a single approach to force employment: two large force packages for two MTWs. By contrast, the new standards provide a wide spectrum of flexible packages. For normal conditions, they disperse forces by creating four medium-sized packages: three for the major overseas CINCs and one held in strategic reserve. For dealing with a single MTW in any one of multiple theaters, they concentrate forces to create a single big and properly tailored package, while maintaining two medium-sized packages for use elsewhere. For dealing with more, different, and bigger wars, they concentrate forces even more, to create two big packages or an even bigger single package. *Their common theme is that they focus on creating appropriate force packages for the full set of purposes and missions ahead, not just for the low-probability event of waging two MTW conflicts at the same time.*

Their intent is to help provide a fresh sense of perspective for judging ways to enhance the U.S. military's flexibility, adaptiveness, and across-the-board performance in the coming era. They will help provide alternative lenses for viewing the strategic priorities of U.S. national security strategy, CINC requirements, service program directions, and force improvement opportunities. They will provide a framework for rewarding investment programs that provide powerful strategic benefits in more ways than one. For example, they will cast a favorable light on measures for creating better infrastructure in new geographic regions, where new bases and facilities might not be needed for future MTWs, but might be essential for new peacetime shaping missions and for responding to small crises and conflicts. They will also help call attention to other attractive measures that may not be given full attention in service programs that today focus on two MTWs.

Like all standards, they must be applied sensibly, with their interplay in mind. Standard 1 should be employed not only for its own purposes, but also to help create adequate capabilities for Standards 2 and 3. Likewise, Standard 2 should be broadly targeted, in ways that have positive effects on the other two standards. Standard 2 calls for being prepared to fight a single regional war, but not only one war in one place. Rather, it means that U.S. forces should be able to wage different kinds of wars, varying in location, strategy, and operations in all three major theaters. The flexible capacity to wage these different kinds of wars will provide an inherent capacity to wage more than one war at a time, should this step become necessary. Standard 3 will no longer rule the roost, but it will still play an important role. It can be used to identify cost-effective measures that help U.S. forces fight not only two wars, but also one war. Examples include strategic mobility, C4ISR systems, war reserve munitions, and stocks: areas where preparing for multiple wars still will make sense.

Together, these three standards will help impart U.S. defense planning with a more comprehensive strategic focus and a better sense of balance. They will put first things first, yet devote proper attention to the full spectrum of critical defense assets. They will help ensure that the still-important requirement of being able to fight two wars does not come at the expense of neglecting other critical measures, especially peacetime operations and strategic shaping missions. Likewise, they will buffer against the reverse risk: that as DoD pays greater attention to Standards 1 and 2, it does not give short-shrift to retaining a two-war capability as a credible insurance policy. By prioritizing this way and allocating funds on the basis of greatest marginal returns, DoD will be better able to build forces that are fully capable of meeting Standards 1 and 2, while still preserving a robust and credible capacity for Standard 3. To be sure, these three standards can be fulfilled only if an adequate defense budget is funded. But these three standards provide an improved strategic formula for gauging force posture and budget requirements, and for determining how priorities can best be established, and scare funds can best be spent, in the event that budgetary shortfalls occur.

In using these three standards, this strategy-based approach will be more complex and harder to explain than the 2 MTW standard, which purchases simple clarity at the expense of strategic sophistication. But this new approach is no more complex than Cold War thinking, which was readily grasped by the Congress and the American people. Within DoD, it would help ensure that defense preparations are targeted not primarily at improbable events, but instead at enabling U.S. forces and CINCs to perform the peacetime and wartime missions that most often must be carried out in today's world. Yes, DoD's Planning, Programming, and Budget System (PPBS) process will be stretched, as will joint operational planning by the Joint Staff and the CINCs.

Preparing the Defense Planning Guidance, Service Program Objective Memoranda (POMs), and CINC operation plans (OPlans) will be harder to accomplish. But surely the Pentagon, in this information age, can carry out defense planning on the basis of three standards, not just one standard—especially since considering all three standards is a good way to determine how to place the defense effort on the right track.

A key payoff of this approach is that senior civilian and military leaders will have significantly better information at their disposal for making the tough decisions facing them. *The bottom line is that this new approach will help create a public rationale that rings true. In addition, it will contribute to creating a better-construed defense effort that supports the full strategic purposes of national security policy and defense strategy. It offers the potential of making the United States and its overseas interests more secure in a still-dangerous and turbulent world where security may be at a premium.*

Conclusion: Keeping Things in Perspective.

What are the implications of this new approach for the future size of U.S. military forces? This question will need to be addressed through careful analysis, but given the directions in which the world seems headed in the coming period, these three standards appear to point to a future posture in the vicinity of today's model, not appreciably larger or smaller. If so, the challenge facing DoD will be one of using similar forces wisely, not building far larger forces or making do with far less. Clearly these three standards do not point toward major force drawdowns: although one standard might require fewer forces than now, the other two standards likely will call for a larger number, similar to now. Indeed, the U.S. Government may ultimately conclude that the act of keeping the peace and shaping the world's future requires more forces than fighting the nation's wars.

In any event, these three standards create no single-point requirement, below which the remaining forces will be clearly inadequate, and above which, added forces are transparently superfluous. But small differences at the margin can make a big difference in strategic performance. If a decision is made to enlarge the force posture somewhat, it likely will focus on adding critical capabilities in areas of deficiency: e.g., "Low Density/High Demand" (LDHD) units, C4ISR assets, logistic support assets, peacekeeping assets, ready reserve component units, and ships and airplanes. A force posture that is 10-15 percent larger than now may be needed, one supported by a defense budget that grows slowly but steadily. But much will depend upon specific assessments of requirements and affordability in the future. At the moment, the important task is to create a sound approach to force-sizing and defense planning, so that such decisions can be made with the best analysis and information available.

The new approach put forth here, with its three standards, is an illustration, not a fixed blueprint. What it helps illustrate is that the past need not be prologue. The existing 2 MTW formula offers one option for navigating the future, but it is not the only viable option. Creative thinking can produce other approaches with attractions of their own. They can be articulated in enough detail to provide concrete guidance for sizing forces, allocating them among missions, and setting sound program priorities for improvements. The challenge is to develop a set of alternative approaches, analyze them, and choose one not because it made sense in the past, but because it offers promise of working best in the future.

ENDNOTES - CHAPTER 3

1. For a current portrayal of the 2 MTW standard and its role in defense planning, see Secretary of Defense William S. Cohen, *Annual Report to the President and Congress, 2000*, Washington DC: U.S. Government Printing Office, 2000, pp. 17-19.

2. For a textbook on planning by multiple objectives, see Ralph F. Keeney and Howard Raiffa, *Decisions with Multiple Objectives: Preferences and Value Tradeoffs,* New York: John Wiley & Sons, 1976.

CHAPTER 4

A NEW STRATEGY AND MILITARY LOGIC FOR THE 21st CENTURY

Huba Wass de Czege
and
Antulio J. Echevarria II

Few strategists would disagree that today's more dynamic strategic environment requires a reevaluation of the ends, ways, and means of U.S. national security and national military strategies. In fact, it has become almost commonplace to declare the Cold War—and its allegedly more stable security environment—at an end. In its place, a "New World Order" has emerged that many argue is rather long on new and short on order. The current force-sizing metric that underpins the national military strategy and, by extension, the national security strategy has of late come under special scrutiny. That metric calls for the U.S. military to maintain the capacity to deter, or if necessary to defeat, large-scale aggression in two major theaters (2 MTW) nearly simultaneously. The military logic for this metric is threat-based and has obvious roots in the Cold War—North Korean and Iraqi threats simply replaced the Soviet one. Architects of the 2 MTW metric have taken great pains to point out that it provides forces to execute other national security missions as well, should either or both of those conflicts fail to occur. Despite these claims, however, experience has shown that the threat-based metric lacks the flexibility required by today's dynamic and uncertain strategic environment. Specifically, the 2 MTW metric does not account for the continuing and diverse requirements of peacetime engagement—shaping the international environment—and it focuses primarily on two geographic regions for crisis response. This essay maintains that the current strategic environment requires a condition-based strategy and a flexible force-sizing metric based on general historical data and validated political-military projections for each specific region.

2 MTW Shortcomings.

The 2 MTW metric does not acknowledge the value of peacetime engagement. Far from mere constabulary duties, peacetime engagement involves actions that facilitate agreements among states and quasi-states, such as Ecuador and Peru, Israel and Egypt, East Timor and Indonesia, and the Kosovars and Serbs, that make crises less likely. It also includes deterring and containing such potential aggressors as Saddam Hussein in Iraq and Kim il Sung in North Korea. In addition, it helps establish a regional balance of power in areas where the potential for trouble exists but the lines of conflict are less clearly drawn, as in the Persian Gulf, the Middle East, and the Taiwan Straits. It supports long-standing arrangements with other national and international agencies to monitor arms control arrangements, counterdrug trade, control the spread of weapons of mass destruction (WMD), monitor embargoes, and so on. It includes responding to humanitarian crises such as those in northern Iraq, Somalia, and Rwanda where military forces can facilitate the work of national, international, private and other nongovernmental agencies. It also helps to ensure that the United States can evacuate American citizens and other noncombatants when hostilities break out unexpectedly. It helps build a basis for cooperative action with regional friends and neighbors that in the long run can reduce the expense of crisis response to the American taxpayer and the burden on American forces. Thus, although it indisputably detracts from some aspects of warfighting preparations, peacetime engagement builds strategically valuable military skills and capabilities that can enable and assist the U.S. military's capacity to wage war.

Unfortunately, the logic underpinning the 2 MTW force-sizing metric wrongly assumes that forces committed to peacetime engagement are available for crisis response elsewhere. Hence, it tends to blur the distinction between forces assigned to the daily work of shaping the international environment and those designated as a hedge against strategic risk. Certain engagement activities—such as peacekeeping missions based on treaty requirements—must continue even during a crisis. Forces engaged in them cannot readily disengage. They become the "sunk cost" of doing strategic business. Many of these forces perform strategically vital functions that can reduce the frequency and escalation potential of crises, or the overall resources required to resolve them. Indeed, U.S. forces in the region help set the conditions for successful crisis response by providing first-hand intelligence, access, and liaison. Although the bulk of those forces would probably remain committed in various locations within a theater, they could still deliver vital deterrent and condition-setting benefits to nearby regions.

Although small-scale contingencies are becoming more frequent, it would be irresponsible to assume that a major war, or large-scale crisis, will not occur. Nonetheless, the 2 MTW metric, associated with counter-aggression scenarios in Southwest and Northeast Asia, fails to account for the unique requirements of other kinds of forceful interventions. Historical and contemporary studies suggest that crises tend to fall into two broad categories. In the first type, as in Panama (1989), immediate action is not necessary and escalation is not likely. Detailed planning and preparations can occur and the response can take the form of a coup d' main, an overwhelming but very focused surprise attack. In the second category, time available for response is relatively short, as in Iraq's invasion of Kuwait (1990), and escalation is possible. Under this category, decisive results might require significant coalition participation to avoid escalation and a rapid and overwhelming military response to prevent the enemy from seizing objectives and establishing integrated defenses.

Force requirements in any crisis have an important mass-velocity dimension that the 2 MTW metric does not capture. Forceful interventions should have an operational momentum that offers U.S. forces a decisive advantage. The rate of force flow often becomes as important as the size, composition, and quality of the force that deploys. The Army, recognizing this requirement, has posted a goal for its Objective Force of being able to deploy five divisions within 30 days as part of a balanced joint force. The product of 2 MTW thinking and logic has produced a condition in which it would take about 75 days to reach this goal. The 45-day gap is a considerable difference, and is not entirely a product of an Army designed to drive to battle from forward deployed garrisons; it is also a function of inadequate strategic transport.

Strategic Alternatives.

Broadly speaking, the United States has three fundamental strategic options as it crosses the threshold into the 21st century. The first of these, which we may call *preventive defense*, aims at preventing problems by deterring, containing, isolating, and defeating specific *threats*. It requires strategists to array forces against such a threat and, thus, amounts to little more than a complex version of the containment strategy that characterized the Cold War. This option pursues a negative end since it seeks to preserve or restore the status quo ante.

The second option, a brand of *neo-isolationism*, pursues a neutral end where security interests are strictly defined in terms of life, liberty, and the pursuit of happiness at home. The evolution of the international environment is largely left to itself. This approach expends less resources in terms of overseas engagement, but it accepts greater risks—and would probably require a larger expenditure of resources—in order to respond to a crisis.

In contrast, the third option, *positive engagement*, aims at establishing a condition of stable peace and widespread prosperity as well as multilateral basis for crisis response. In other words, its ends are *positive*, rather than negative or neutral. A strategy of positive engagement accords well with a strategic environment in which vital interests are defined by conditions (peace, freedom, rule of law, and growing prosperity) rather than by the containment or defeat of inimical state or non-state actors. Simply stated, the basic aim of positive engagement is to maintain and shape the peace with allies and partners who share—or at least are not opposed to—our interests and core values.

Like the other options, a strategy of positive engagement depends on the coordinated use of all forms of national power. However, it differs from the other two, and with the current policy of Engagement, in one important respect: it calls for proactive action—seizing the initiative—in global security affairs to widen the circle of stakeholders in global *peace and prosperity and to set better terms when responses are required.* It also differs in that positive aims generally require more energy and resources than negative or neutral ones. Not only would positive engagement reduce the frequency of international crises, it would respond strongly enough so that the post-crisis condition amounts to a more stable peace than the status quo ante. Positive engagement would thus require a new approach to national military strategy as well as a redress of the current imbalance in U.S. military power.

Accordingly, the national military strategy that flows from positive engagement would focus on establishing and expanding regional "zones of security" characterized by multiple systems for facilitating crisis response and conflict containment, as well as the entire range of stability and support operations. Military presence (both in and out of region) must signal a definite commitment to deterring aggression and must demonstrate a bona fide capability to respond decisively to conditions that threaten the interests of the United States and its fellow stakeholders. It goes without saying that under this strategy (and others), U.S. military forces would have to achieve and maintain the capacity for swift, decisive, and focused intervention.

However, the underlying logic that separates this strategic approach from the others is that military forces must have the capability to take proactive action in peace, not just react to crises. In other words, the entire defense establishment would have to think differently about what military forces can or should do in peacetime. For example, the U.S. military's geographical Commanders in Chief (CINCs) might conduct proactive regional engagement campaigns on a daily basis. These operations would require a greater number of flexible and regionally "street-smart" forward-presence conventional and special forces.

Positive engagement and its supporting military strategy would also require the defense establishment to change the way it approaches crisis response, contingency operations, and major theater wars. The best efforts to prevent war will not necessarily preclude it. To be sure, not every crisis will require a military solution. However, maintaining global peace and prosperity will sometimes necessitate a military response. That response must come from an understanding of the conditions and forces that were at work during the peace and those necessary to produce a more stable peace at the resolution of the crisis. More specifically, the defense establishment will have to alter its assumptions about the long and short-term effectiveness of its demonstrated preference for stand-off, primarily technological military responses. As the war in Kosovo demonstrated, such one-dimensional responses permit an adversary to prolong the fight, carry out his political objectives, and put stress on the "glue" that holds the opposing coalition together in the process. More to the point, military actions aimed at defeating and removing adversarial regimes will require a more balanced—full dimensional—U.S. force than current defense assumptions acknowledge. Indeed, given the speed with which regional crises can unfold, and the strength an adversary can muster locally, a decisive force must respond as soon as it is evident that war will ensue. The length of the campaign, decisiveness of the results, and degrees of collateral damage will be a function of the combined strength, balance, and tempo of the response.

A New Metric.

The force-sizing metric that supports positive engagement should, first of all, allow for a significant resource investment in peacetime engagement. The size of a CINC's baseline force, both stationed abroad and in CONUS, should reflect the "work load" history of the previous 5 years and expectations for the next 5 years. If the events of the last 5 years are any indication, these requirements will probably increase rather than decline. U.S. force requirements under a strategy of positive engagement might, for example, include continued participation in the Balkans as part of a NATO task force but augmented by a sub-regional engagement effort aimed at building long-term cooperation under the Organization for Security Cooperation in Europe (OSCE). In the Middle East, the United States might contribute elements to a standing United Nations (U.N.) peacekeeping force as well as a larger subregional engagement effort involving key G-7 and NATO allies. In the African subregion, U.S. forces might find themselves suddenly conducting noncombatant evacuations, disaster relief, and stability operations, either unilaterally or in conjunction with other stakeholder nations. The point here is that historical data combined with validated political-military projections—with a hedge against uncertainty—would serve as a basis for justifying regional force requirements for upcoming program objective memorandum (POM) years.

As previously mentioned, peacetime engagement requirements do not disappear during a crisis and may in fact increase in importance. Hence, the new force-sizing metric must include adequate forces for responding to a number of crises. The first step in determining what size force is adequate for crisis response is to identify the potential number of crises that might erupt and how many of them possess a real potential for escalation. Second, the metric must include sufficient resources to permit U.S. forces to achieve and maintain a decisive operational momentum.

It is beyond the scope of this chapter to offer specific numbers of forces. For that, the defense community must turn to historical data and regional political-military projections and analytical organizations, such as the Center for Army Analysis, that can help quantify the force requirements. For purposes of illustration, we suggest that the U.S. Army's regional commitment to positive engagement consist of a baseline force consisting of a theater army, Army special operations forces, a forward-presence corps, and a corps equivalent of reserve component forces for rotational employment. This baseline force would consist of active and reserve component forces. Active components, rounded out with reserve component elements, would conduct the daily work of peacetime engagement. Concurrently, a corps equivalent of reserve component forces would conduct long-range planning for full integration into the recurring work of peacetime engagement. As a rule of thumb, then, reserve forces would be allocated against the more predictable requirements of the future, and active forces against initial commitments and those most likely to change.

In a dynamic and unpredictable environment, a military force capable of rapid and decisive response becomes a prerequisite rather than an option. These forces should be separate from those committed to peace engagement. The national command authorities should maintain a rapid crisis response force with broad, tailorable capabilities to intervene with *full-dimensional* force in cases involving combinations of small states and the non-state actors associated with them. In addition, they should maintain a strategic response force capable of backing up early deploying forces when the response involves medium powers and combinations of conventional and unconventional forces.

The legacies of the Cold War continue to influence U.S. strategic thinking. The defense community continues to justify its strategic preferences with *threat*–based assessments, some of which are loosely labeled *asymmetric*. However, a threat-based strategy has serious liabilities in an environment in which the next opponent or the next crisis is nearly impossible to identify. Today's strategists might learn something from law enforcement agencies and fire fighters who do indeed attempt to track and apprehend "America's Most Wanted" criminals, but in fact focus primarily on enforcing the law and maintaining a condition of freedom and safety. Similarly, today's strategists ought not to cease tracking specific threats entirely, but should concentrate on how the United States can establish and maintain a long-term condition of global peace and prosperity.

CHAPTER 5

RETHINKING TWO-WAR STRATEGIES[1]

Michael E. O'Hanlon

As new administration officials focus on the next Quadrennial Defense Review (QDR) in January 2001, they should rethink the two-war construct. While some sort of multi-war capability is needed, the notion of two DESERT STORMs has outlived its usefulness.

It is not hard to find critics of the two-DESERT STORM approach which had its origins in efforts by Secretary of Defense Richard Cheney and General Colin Powell to design a post-Cold War base force and gained more popularity under the Clinton administration during the Bottom-Up Review in 1993 and the last QDR in 1997. However, few have proposed an alternative approach. Specificity is both needed and overdue. Replacing the two-DESERT STORM paradigm with a concept for force-sizing that could be called DESERT STORM-plus-DESERT SHIELD-plus-Bosnia (IFOR) warrants consideration. Though the term may be cumbersome; after a decade of the two-DESERT STORM jingle we have oversimplified force planning long enough.

This new approach might allow further modest personnel reductions. But its main effects would be on the structure, not the size, of the Armed Forces. Specifically, it would permit a force posture more conducive to executing the types of missions that have recently strained the military. The reasons why it would not jeopardize core national interests are developed below.

Out with the Old.

The congressionally-mandated report released by the National Defense Panel (NDP), which was published 6 months after the QDR report, concluded that the two-theater war construct has been a useful mechanism for determining what forces to retain as the Cold War came to a close. But it is fast becoming an inhibitor to reaching the capabilities we will need in the 2010-2020 time frame. The panel regarded the two-DESERT STORM concept as little more than a bureaucratic device that was more relevant to institutional requirements than to real world threats.

However, the dismissive view of the NDP position went too far. Saddam Hussein and Kim Jong-Il continue to threaten U.S. interests. We cannot drop the two-war construct until convinced that any successor concept will afford adequate deterrent and defense capabilities. Vague musings by the panel about the two-war framework, though useful as cover for debating this subject, hardly form the basis of a new national military strategy.

The way in which the panel dismissed the two-war approach provided Secretary of Defense William Cohen with an easy comeback: which threat should be ignored, Iraq or North Korea? And which national interest should be abandoned, ensuring access to Persian Gulf oil or maintaining the security of South Korea (not to mention general stability and nonproliferation in both theaters)? As long as critics of the two-war framework propose replacing it with a single war capability, they will lose the force planning debate to such forthright rebuttals. The ability to handle overlapping crises in two or more locations is indeed a sound strategic pillar on which to base U.S. forces.

In a broader sense, however, the NDP report was right. Positing two simultaneous replays of Operation DESERT STORM, most likely in the Persian Gulf and Korea, smacks of preparing to refight the last war. Moreover, it presupposes that we would use virtually identical types and numbers of forces—in each case six to seven active-duty ground combat divisions including Army and Marine Corps contributions, additional ground combat units from the Reserve Components, ten wings of aircraft, four to five carrier battle groups, and other assets. Whether operating on the open desert of Arabia or Bosnia-like terrain in Korea, and whether supported by relatively weak allies in the Persian Gulf or the capable forces of South Korea, planning documents call for roughly the same cookie-cutter U.S. force package: a slightly smaller version of that which fought Operation DESERT STORM.

If there were no opportunity costs to keeping the two-DESERT STORM planning framework, the Pentagon would suffer little harm in retaining it. But given likely fiscal constraints in coming years, keeping a high-priced insurance policy against regional conflict would make it impossible to afford other key defense investments and thus would leave the Nation vulnerable on other fronts. It would also leave us with a force structure not well suited to smaller operations meaning that ongoing no-fly-zone missions and peace operations will continue to overwork our personnel.

The United States should change its warfighting strategy from the two-war concept to what can be called a DESERT STORM-plus-DESERT SHIELD approach. A force of 200,000 troops was sent to protect Saudi Arabia during Operation DESERT SHIELD in 1990. By contrast Operation DESERT STORM employed 500,000 American troops to oust Iraq from Kuwait. Actually, it would be more accurate, if more unwieldy, to term this approach a DESERT STORM-plus-DESERT SHIELD-plus-Bosnia (IFOR) strategy. The latter two need not be seen as simultaneous all-out conflicts because, at some point, worst case analysis must be plausible. But the requirement to maintain deterrence and presence, while waging a Operation DESERT STORM-like operation along with something akin to Operation DESERT SHIELD, seems compelling. This type of construct would still be somewhat artificial, but it would encompass a fuller and broader range of likely U.S. military missions than the current planning framework.

The alternative would still require 90 to 95 percent as many active duty personnel as current plans. The Operation DESERT STORM package would have to err on the side of caution, including a cushion of extra forces in the event the United States and its allies encountered unexpected difficulties such as widespread enemy use of weapons of mass destruction. For example, it might require a total of six Army divisions and twelve Air Force fighter wings as well as currently anticipated levels of Navy and Marine Corps assets. Backup exists in the Army National Guard, which retains almost as much of the combat force structure as the active Army but would have been expected to deploy less than 20 percent of its units into combat under the 1997 version of the two-DESERT STORM plan. Adding a division for a major peace operation would leave an active duty Army perhaps 90 percent as big as current levels, with slightly smaller cuts in other services.

Something Has To Give.

But in a period of fiscal surplus, why not keep the two-war capability while simply adding more forces as needed? The budget situation is admittedly less stark than it appeared at the time the last QDR even though readiness costs have also grown, laying claim to part of the Department of Defense (DoD) share of the budget surplus. Overall, rosy forecasts notwithstanding, it is doubtful that the military will be able to retain current force structure and modernization programs. Large cuts will not be needed, but trimming probably will be.

Budget plans substantially increase procurement for two reasons. First, the spending spree of the 1990s must end because systems purchased during the Reagan era are wearing out quickly. Second, the Pentagon intends to replace existing weapons with more expensive ones like F-22s, not to mention joint strike fighters and F/A-18E/Fs, improved attack helicopters, and submarines. The belief appears to be that increasing procurement to $70 billion per year from the 2001 level of $60 billion will pay anticipated bills. But neutral watchdogs like the Congressional Budget Office tend to estimate steady-state price tags of $80-90 billion for the future force in constant 2000 dollars.[2]

Meanwhile, other budgetary demands are likely to hold steady or rise under existing plans. Personnel spending will no longer decline because real pay raises will more than counter savings in personnel still to be made in the final stages of the post-Cold War drawdown. Though some hope to realize large savings through privatizing and outsourcing as well as base closings, particularly in operations and maintenance, savings will be modest. Health care, maintenance, and base cleanup continue to exert upward pressure on the budget. Meanwhile reductions in research, development, test, and evaluation are being questioned as unwise and would not save much.

The bottom line is that real defense spending will likely have to grow by at least $30 billion in the decade ahead to sustain the current force and planned modernization agenda. In other words, spending must increase from the 2000-2001 levels of around $290-320 billion or more. With an available surplus nearing $2 trillion projected for 2001-2010 (not counting surpluses in Social Security and Medicare), that may not seem to be an inordinate defense spending increase because it would probably total only about 20 percent of available funds.

However, expecting the Pentagon to get $300-500 billion in the next decade is highly optimistic. Out of a $1.8 trillion projected surplus, $600 million would be needed to preserve existing levels of domestic services and allow spending to grow as fast as population rather than just keeping up with inflation. Because many discretionary spending programs in transportation, education, immigration, prisons, and environment are linked to the size of the population or economy, that is a prudent assumption. Efforts to shore up entitlement programs in the long term, given high priority by both political parties and presidential candidates, are likely to require at least $500 billion over the next decade, according to the Center on Budget and Policy Priorities. That leaves $700 billion for tax cuts as well as prescription drug benefits for the elderly and education. After all is said and done, it is highly unlikely that anything close to half a trillion dollars in real funding will be added to the DOD budget over the next decade.[3]

The gap between planned outlays and likely resource levels for defense is likely to amount to $10-20 billion per year over the next 10 years. Part of the gap can be closed by reducing service modernization agendas. Absent competitors and given advances in computers, electronics, and robotics, less emphasis should be put on extremely expensive weapons platforms and more on a system-of-systems approach. But even such a radical change in acquisition may not solve all budgetary problems. That means that a modification of the two-war strategy (as well as cutbacks in nuclear forces and a willingness to try new ways to maintain forward presence in the Navy and Marine Corps) is likely to be a budgetary imperative.

Even more important, altering the two-DESERT STORM construct is necessary for the well-being of the Armed Forces. Adapting a less demanding two-war capability would allow the Army to shift personnel from traditional combat roles to the types of low-density/high demand support activities that are typically overused in today's non-warfighting missions.

A Rapidly Deployable Force.

A 200,000-strong DESERT SHIELD force would be extremely effective. If deployed promptly, it could defend allied territory and infrastructure against virtually any threat on the horizon today. U.S. commanders were confident that they could defend Saudi Arabia with a DESERT SHIELD force in 1990. Today the high caliber of personnel, combat equipment, and support capabilities such as advanced reconnaissance systems would make such a DESERT SHIELD capability significantly superior to the notional regional aggressor force specified in the Bottom-Up Review, even though the latter force might be two to three times larger.

The airpower component of a DESERT SHIELD-like deployment, smaller but about as capable as that of DESERT STORM and larger than that deployed against Serbia during Operation ALLIED FORCE in 1999, could devastate enemy forces and industrial infrastructure. The ground component could conduct certain offensive land operations. General Norman Schwarzkopf considered evicting Iraq from Kuwait with a force of this size before asking Washington to double the deployment (and that was before improvements made the military better armed than a decade ago).[4]

The odds that such a force could deploy in time to prevent significant loss of territory are reasonably good. Since the Cold War, the Armed Forces have positioned more equipment abroad and bought more fast sealift in the form of large medium-speed roll-on/roll-off ships. In addition to forces routinely deployed overseas, including 37,000 in Korea, somewhat more in Japan, and half as many in the Persian Gulf, Army brigade sets of equipment are based in Kuwait and Korea, another is afloat off Diego Garcia, and elements of a fourth are in Qatar. Marine brigade-equivalent sets are at sea at Diego Garcia and Guam and in the Mediterranean. These units could be married to troops from the United States in a week or so. Further improvements in both lift and prepositioning could shorten response time for other units too. Just as important, stocks of precision guided munitions are now located overseas. Stopping an enemy quickly and hitting it from the air might make a major ground counteroffensive unnecessary. At a minimum, it should reduce its urgency.

Hollowing Threats.

The militaries of Iraq and North Korea remain dangerous but are markedly weaker than several years ago. Moreover, neither power is likely to get much stronger any time soon. This increases the odds that the United States with a DESERT SHIELD force, and its regional partners, could prevent significant loss of allied territory. Iraqi conventional forces remain about half the size and strength of 1990. As opposed to a pre-DESERT STORM inventory of 5,500 tanks, Baghdad now has 2,200. Levels of light tanks and armored personnel carriers are down from 7,500 to 3,000; troop levels have declined from 1,000,000 to 400,000.[5]

The Defense Intelligence Agency reported in 1997 that although North Korean forces are poised near Seoul, their capability to conduct large-scale combat operations continues to deteriorate as worsening internal economic conditions undermine training, readiness, and sustainment. And subsequent threat assessments reconfirm that decline, notwithstanding some modest improvements reported in readiness levels over the last year.

To be sure, South Korea remains vulnerable to artillery, missiles, and special forces from the North, and Pyongyang unquestionably possesses what amounts to massive terrorist assets to target against Seoul. Any war on the peninsula would cause untold civilian deaths as well as large numbers of military casualties. But there is a difference between terrorism and an invasion.

Indeed, the Iraqi and North Korean threats have declined enough that 200,000 to 300,000 U.S. troops might even suffice for a counterattack. A single robust DESERT STORM-like capability of closer to half a million troops should be retained out of prudence. But there is less and less reason to think such a large force would be needed even for a march on Baghdad or Pyongyang.

Allies Count.

Economic troubles notwithstanding, the South Korean military is improving and, together with modest American forces in place on the peninsula, could probably withstand an attack. South Korea, combined with U.S. forces the 2d Infantry Division and forward-based airpower could inflict great damage to North Korean forces and could most likely stop an assault well short of Seoul. At a minimum, they could buy enough time for U.S. reinforcements to arrive.

Most military casualties would be North Korean. Its military is more obsolescent than that of Iraq; and any invasion attempt would have to cross the most militarized swath of ground on the planet. The density of forward-deployed allied forces near the demilitarized zone (DMZ) is greater than was the density of NATO troops along the intra-German border during the Cold War. North Korea would have to rely on roads and bridges that would surely be destroyed in the first minutes of combat. If attacking near Seoul through the Chorwon or Munsan corridors, the invaders would have to cross the Han or Imjin Rivers. Both freeze in the winter, but the ice might not be strong enough to support a large armored force. North Korean chemical weapons, commandos deploying through tunnels, and forward-deployed dug-in artillery would complicate the battle and cause many casualties. But armor would have great difficulty breaking through allied lines and reaching Seoul.

Although the South possesses less armor than the North, its technological edge evens the balance of tanks, artillery, planes, and other heavy equipment according to some assessments. Its armor is nearly equal that of U.S. models; for example, the K-1 tank is based on the M-1 and uses some of its important components.

Given the higher state of military readiness of South Korea, it is reasonable to conclude that its forces are superior to those of the North. Looking at the outcomes of a range of past battles, one analyst estimated that such readiness factors can at least double combat capability. Despite the fact that, as another analyst pointed out, DoD models appear to assume that South Koreans would not fight as well as North Koreans, the former are competent soldiers and extremely well postured to stop an invasion.[6] An attacker attempting to directly penetrate densely prepared positions usually advances only a couple of kilometers a day even when not outclassed technologically, as the North Koreans certainly are. Given the lethality of modern airpower and U.S. ability to quickly fly in combat jet reinforcements, such a slow pace of advance itself generous to the North would be a recipe for disaster on the part of an invasion force.

Pyongyang could not pull off a left hook or bypass the Korean equivalent of the Maginot Line because the defenses extend across the peninsula. In addition, the allies enjoy overwhelming dominance in all-weather, day/night reconnaissance that watches over all significant movements. But chemical and biological weapons pose a special threat, especially given the limited confines. U.S. forces have increased attention to such threats, with the QDR initiatives raised by former Secretary Cohen being especially noteworthy. One could argue that Seoul should do more as well. But it is more difficult to employ chemical weapons than is commonly asserted, especially for an infantry force like North Korea's. For example, it is extremely challenging for a foot soldier, suited up in bulky and probably rather substandard protective gear, to cover many kilometers to take advantage of holes in enemy lines created by chemical attack. Nor should the North blithely assume that such attack would not be countered by U.S. nuclear retaliation. Airbursts in corridors north of the demilitarized zone would cause little harm to friendly forces while considerably affecting North Korean units. They would also send a powerful message that America will not tolerate the employment of weapons of mass destruction against its troops or those of allies.

There is a final argument against the two-war construct. Just as the capabilities of South Korean forces must not be ignored, one should not overlook the likely role that British forces would play in a conflict in the Persian Gulf. The United Kingdom deployed 30,000 troops during DESERT STORM, was prepared to send 50,000 troops to fight against Serbia, and tends to be aligned with the United States on issues of war and peace in Southwest Asia.

Without prejudging the prospects for an integrated European military force, or presuming full agreement between Washington and London in matters of defense and foreign policy, one can venture to say that Britain would probably provide a division and several fighter squadrons to any coalition led by the United States in a future conflict in the Persian Gulf. However, pessimistic American war plans do not now assume such contributions.

Some will see the similarity between this proposal and a plan put forth as a trial balloon by Secretary of Defense Les Aspin in 1993. Known as a win-hold-win strategy, it envisioned completing an all-out war in one theater while simply holding the line in another. Once the first war was won, forces would be redeployed for a counteroffensive to meet the other challenge. But the caricature of that approach understated its capabilities and doomed it to rejection. Derided as win-hold-oops because of its alleged risk to war plans, it never stood a chance bureaucratically or politically.

The important point is that a DESERT SHIELD force, with its overwhelming airpower and other long-range strike systems, can do more than hold a defensive line despite the limited capabilities of such a force.

The next QDR should weigh arguments like those outlined above. The alternative is attempting to prevail in simultaneous worst-case scenarios in the Persian Gulf and Korea (something that the Armed Forces could not have handled even during the Cold War, given U.S. commitments in Europe) at the expense of readiness, research, and preparing for the future. More dangerously, the military could continue to overuse and wear out its most precious assets and its people. That would be a far greater risk than the remote possibility of two nearly simultaneous, all-out conflicts against both Iraq and North Korea.

ENDNOTES - CHAPTER 5

1. This chapter first appeared in *Joint Force Quarterly*, Spring 2000, pp. 11-17.It is reprinted with the permission of the author and of *Joint Force Quarterly*.

2. The Congressional Budget Office has estimated that DOD may be required to spend $90 billion annually on procurement, given current plans combined with the need to replace certain systems that the services have not yet incorporated into their formal acquisition programs. See testimony by Lane Pierrot, "Aging Military Equipment," Subcommittee on Military Procurement, House Committee on Armed Services, February 24, 1999.

3. Robert D. Reischauer, "The Dawning of a New Era," in Henry J. Aaron and Robert D. Reischauer, eds., *Setting National Priorities*, Washington: Brookings Institution, 1999, pp. 1-12.

4. See Michael R. Gordon and Bernard E. Trainor, *The Generals War: The Inside Story of the Conflict in the Gulf*, Boston: Little, Brown, 1995, pp. 123-41; Robert H. Scales, Jr., *Certain Victory: The U.S. Army in the Gulf War*, Washington: Brassey's, 1994, pp. 121-128.

5. International Institute for Strategic Studies, *The Military Balance 1990-1991*, Oxford: Brassey's, 1990, p. 105; *The Military Balance 1997/98*, Oxford: Oxford University Press, 1997, p. 127; *The Military Balance 1999/2000*, Oxford: Oxford University Press, 1999, p. 134.

6. Trevor N. Dupuy, *Attrition: Forecasting Battle Casualties and Equipment Losses in Modern War*, Fairfax, VA: Hero Books, 1990, pp. 105-110, 148; Lawrence J. Korb, "Our Overstuffed Armed Forces," *Foreign Affairs*, Vol. 74, No. 6, November/December 1995, p. 25.

CHAPTER 6

WHAT FOLLOWS THE 2 MTW FORCE SHAPING PARADIGM?

Daniel Goure

It is fast becoming the conventional wisdom that the two Major Theater War (MTW) standard for the size and character of U.S. Armed Forces is dead. The reasons given for this judgment are various; there are not two enemies to confront the United States; there are insufficient resources to support a force as large and complex as that required by the standard; and, the standard does not reflect the current realities of peacekeeping and small-scale contingencies. It has been evident for several years now that U.S. military forces are being over-stretched and over-used. It is not possible for the military to both maintain the capability to fight two nearly simultaneous MTWs and also support contingencies from Somalia and Haiti to the Balkans and the Persian Gulf. The divisions supporting Balkans deployment have been rated as C4, or unready for wartime duty precisely because they are engaged in peacekeeping duties. The present force structure lacks sufficient low-density/high-demand assets to support adequately the current set of demands. The campaign in Kosovo, when operations over northern Iraq were curtailed in order to reinforce forces conducting the air war, demonstrated that there are not enough of these units for more than one MTW.

Despite the many criticisms of the two MTW standard, it is good to remember that it was directly tied to America's unique role and position in the world as well as to a national security strategy which was one of managing the international system to prevent conflicts from arising. This point was made rather well by Secretary of Defense William Cohen:

As a global power with worldwide interests, it is imperative that the United States now and for the foreseeable future be able to deter and defeat larger-scale, cross-border aggression in two distant theaters in overlapping time frames, preferably in concert with regional allies. Maintaining this core capability is central to credibly deterring opportunism—that is, to avoid a situation in which an aggressor in one region might be tempted to take advantage when U.S. forces are heavily committed elsewhere—and to ensuring that the United States has sufficient military capability to deter or defeat aggression by an adversary that is larger, or under circumstances that are more difficult, than expected. This is particularly important in a highly dynamic and uncertain security environment.[1]

Abandoning the commitment to two MTWs does not mean rejecting the principle that the primary role of U.S. military power is to fight and win this country's wars. But what wars will these be? Some analysts have argued that the problem is not one of focusing the military on the need to fight wars but that the current standard is overly specific, focusing on a single type of conflict. One of the earliest critics of the two MTW standard was the 1997 National Defense Panel (NDP). The NDP characterized the two MTW construct as "a force sizing function and not a strategy." The NDP went on to warn that should one of the major contingencies disappear, particularly in the absence of an alternative strategy, that "a strong demand would be created for deep and unwise cuts in force structure and personnel."[2]

The simple solution, to trim the force structure at the edges, is one that would fall into the trap pointed out by the NDP. This would produce a "one and a half MTW" standard or, perhaps, a "one MTW plus two or three smaller-scale contingencies." Such an approach would not change the basic strategic focus of U.S. national security strategy and force development, but could reduce the carrying costs for what some critics view as an overly large, too expensive force posture. However, it would also reduce any cushion the U.S. military would have in the event that the one MTW was larger or more difficult than expected. Military planners would have a natural bias towards hedging on the character of the one MTW force. A single MTW against a more robust adversary could require nearly as large a force structure as has been deemed necessary at present to meet the canonical two MTW threat.

In addition, the size and composition of the present force is driven as much by peacetime demands as by war fighting requirements. Former Chief of Naval Operations Admiral Johnson testified recently that peacetime deployment requirements were the true sizing metric for the Navy. In fact, a recent Navy force structure study argued that more surface combatants were required to meet projected operational tempo (OPTEMPO). The Air Force's current force structure problems, particularly for tankers, transports, and low-density/high demand assets, is also a function of the increased peacetime OPTEMPO. It is possible to change from the two MTW standard to something ostensibly less stressing and not shave much off the force.

But, if the two MTW standard is dead, what should will replace it? The most likely solution, noted above, is far from satisfying. Moreover, the "two MTWs Lite" approach really only addresses the question "How much is enough?" These simple alternatives do not address the overriding question: "What strategy is the new force-sizing paradigm intended to support?"

The majority of discussions of alternatives to the two MTW standard have focused on a very narrow set of political assumptions. They would change the emphasis in U.S. planning, but few of the basic assumptions. They tend to be threat-driven, or in some cases lack of threat driven. Nevertheless, they provide no conceptual alternatives to the current paradigm, merely a different way of assessing how much is enough.

Perhaps it is would be more useful to define the principles that should animate the search for an alternative formulation. There are a number of potential alternative approaches for defining both the size and quality of a future force posture. They can be classified generally as either top-down or bottom-up. The former consists of approaches that are derived from alternative strategic concepts. Different ways of defining national interests, the role of military force in security policy, and the relationship between U.S. security and that of others, could result in distinct and different requirements for military forces. The latter group would be those approaches based less on such questions as what we might to fight about and against whom and look instead to changes in the means and methods of warfare to build alternative force structure paradigms.

Top-down Approaches.

The first alternative strategic paradigm shift, one which would most significantly shift force planning away from the current war fighting model, would be one focused on collective security. This approach would extend the argument that ours is a time without the threat of conflict between the major powers. The source of the majority of significant wars of the last 500 years, European inter-state politics, has been tamed by the twin forces of military dependence on the United States and economic integration. War, according to this view, is no longer an option between major powers. The forces for change in the 21st century world, catalogued in Thomas Friedman's book, *The Lexus and the Olive Tree*, which include globalization, greater access to media, and improved education, to name three important variables, will reduce the economic, political and social incentives to engage in major conflicts. To these forces could be added the effect of changing demographics. The first world is aging rapidly. Older populations are less likely to make war. The third world, where the bulk of the population growth is occurring, will be overwhelmed by the problems of modernization, and will have little in the way of resources, human or capital, to devote to serious war-making capabilities.

The uses for military force would be significantly restricted under this paradigm. The role of force would be to create and maintain stability. Shaping missions would become the first priority. Crisis response, including responding to humanitarian events, so-called failed states, and the possible rise of hostile non-state forces such as terrorist organizations would be the second priority. Monitoring would become the third priority role.

There would be a role for traditional forces to support the crisis response mission. Such a force could reflect the European Union's concept for its own rapid reaction capability. The United States would deploy a rapid-reaction corps, plus supporting air and ground elements. The bulk of U.S. forces would focus on the shaping and monitoring missions. They would be light forces, with an increase in the number, and even types, of so-called low-density/high-value elements.

A very different force alternative would be created by a paradigm that emphasized continuing the current U.S. global preeminence. Some have characterized this approach as "empire maintenance."[3] This concept would view major inter-state conflict as unlikely. The United States would be required to identify a security area in which it would apply the principles of empire management. It would be expected that other powers would do so, too. Overall stability would come through a concert of these major powers. The principal sources of danger would be from local conflicts that got out of hand and from lesser threats to the "imperial order." However, conflicts outside the boundaries of the empire, would not be a U.S. concern in this paradigm. The United States would also have to be rather circumspect in its judgments of the ways other great powers manage their imperial spaces. The United States reaction to the Russian war in Chechnya suggests that elements of this model already are in place.

This model would differ from classic balance of power, however. It assumes that the goal of all the powers is system stability. As a result, the role of military power would not be deterrence of other powers, but rather the prevention or control of local conflicts. Forces would need to be forward deployed, perhaps on the model of the Roman legions, serving both a constabulary role and as a deterrent to would-be local aggressors. It could be argued that the U.S. military presence in Europe is operating according to this model.

This paradigm could suggest a U.S. force structure oriented around a set of forward-deployed joint task forces, each capable of addressing the threats in their own area of concern. There would be a small, strategically-deployable reserve capability. Greater reliance would be placed imperial auxiliaries—that is, allies. Local allies would both supplement U.S. capabilities and undertake independent missions. U.S. forward-deployed forces might be oriented around high-end military capabilities, while allies provided other, unique capabilities.

A third model would be that of classic Realpolitik. While the times and technology change, the nature of politics does not. Nations will still employ military power and engage in conflicts, in pursuit of their political goals. This model would assume that the danger of major conflict between the powers continues to exist. It might be argued that this danger is likely to grow as new power centers emerge, the United States "unipolar" moment passes, and nations pursue their particular and individual interests. Other great and emerging powers, Russia, China, Japan, India, the EU, would be the principal focus of strategic concerns. Conflicts would be caused mainly by the efforts of these nations either to increase or resist erosion of their relative power position vis-à-vis other powers. The source of a particular conflict could be political, economic, or geographic in character.

The Realpolitik paradigm would not assume a return to a Cold War-like environment. Rather, it would be much like the world between the last quarter of the 19th century and the first decade of the 20th century. The focus of security concerns in this paradigm would be the relative power relationships of the major powers on the Eurasian landmass. That is the region in which all the powers, save the United States, reside. Alliance relationships would clearly become of greater significance in this environment than has been the case since the end of the Cold War. Specifically, this environment could lead to the re-establishment of NATO as the essential collective defense organization for the West. The power struggles would occur along the margins, in the spaces between these powers. Regions of concern would include the Caucuses, Central Asia, the Middle East, Eastern Europe, and possibly the Koreas.

It is not clear that even a robust two MTW standard would be sufficient in this environment. The purpose of U.S. forces would be to deter major conventional war and to prevent escalation to nuclear use. Forces would be designed to fight and win large-scale conventional conflicts. Planning would have to consider the idea of protracted conflicts. Significant reserve capabilities would be required.

The role of strategic forces would have to be rethought. Their use could be restricted to limiting the scale and destructiveness of future conflicts. In effect, strategic forces would be designed to keep conflicts between the major powers limited. Strategic defenses could well become a vital investment. Reductions in strategic nuclear forces could still proceed, although tactical nuclear weapons would probably need to be reintroduced into the arsenal.

The last strategic paradigm could be termed "Defensive Independence." The central assumptions of this paradigm are: that the United States faces no peer competitors or significant regional adversaries, that the United States continues to be *primus inter pares* in the global economy, and that domestic sentiment tends to move the United States away from ideas of shaping environments and humanitarian responsibilities in areas not of vital interest.

Regional allies and coalitions take on the responsibilities to manage local instabilities. The most significant security problem for the United States is to deny any potential rogue state the ability to upset the stability or threaten the United States.

A paradigm based on "Defensive Independence" would focus security policy on a limited number of regions or states of vital interest. So long as they were not under threat, the United States would remain essentially aloof from local security issues. In addition, the United States would require that the homeland be secured against threat. Defensive, vice offensive, measures would be preferred as a means of protecting national security. Defense policy would seek to deny potential adversaries the means to threaten access to regions and states of interest, or to threaten the homeland.

The critical defense policy concerns under this paradigm would be continuity of U.S. access to those regions and states of interest, denial of direct attack options on those regions and states by rogues, and ensuring the protection of the homeland. Naval forces and long-range strike capabilities would hold the dominant place in the force structure to support this paradigm. The United States could seek to rely more heavily than at present on nuclear weapons to deter threats to vital interests. A stronger emphasis on strategic defenses would also be warranted.

Bottom-up Approaches.

The bottom-up approaches would take as given the current security environment and the present U.S. role in it. Alternative paradigms would be based on the ways forces could be built to achieve essential core objectives, without the need to define either future war waging requirements or conflict scenarios. As a result, bottom-up approaches have relatively little to say about alternative force sizing concepts.

The first of the bottom-up paradigm might be termed "Technology-Driven." A "Technology-Driven" paradigm would seek to make use of the so-called Revolution in Military Affairs (RMA). In its simplest incarnation, this paradigm would assume that military superiority would be conferred in that nation which most rapidly identified and best exploited the technologies at the core of the RMA. U.S. military strategy, doctrine, and force shaping criteria would be based on the exploitation of technologies in which the United States holds an asymmetric advantage over potential adversaries. The United States would seek to fight the types of conflicts or in ways that reflected the exploitation of technological advantages. In theory, this approach would require a focus in force planning and acquisition on the maintenance of high-end conflict capabilities.

An example of an attempt to define a "Technology-Driven" paradigm is the Information Warfare (IW) debates of the early and mid 1990s. In these debates, proponents of the IW school suggested that a new mode of warfare was emerging based on the ability to control and manipulate information. So-called kinetic warfare, the use of bombs and bullets, was viewed as an increasingly antiquated capability. The "Technology-Driven" paradigm looks for transcendent warfare models, technologies that are revolutionary not simply at the level of military forces and doctrine, but at the operational and strategic level.

Arguably, a "Technology-Driven" paradigm would lead to an emphasis on aerospace and information capabilities in force design. There is a clear U.S. military advantage in these areas. Each of the Services is intensifying its investments in advanced aerospace/IT capabilities. The costs involved in pursuing such an option would be high and require an "unbalancing" of the force structure relative to the current posture.

It must be admitted that a "Technology-Driven" paradigm based on an advantage in aerospace capabilities would not be able to deal fully with the entire range of security problems that currently confront defense planners. The choice of an aerospace-centered focus for defense planning and force shaping would require heavier reliance on allies to fill capability gaps, specifically in ground forces, created by the U.S. force shaping choice.

A second bottom-up approach could be called "Capabilities-Driven." This paradigm would not use any particular set of scenarios to determine force sizing and shaping goals, but define a set of overarching capabilities that would be the basis for force design. Desired capabilities could be defined relative to threat characteristics or on the basis of canonical objectives. For example, a desired capability might be to find and destroy a certain number of mobile targets in a given time period. Another might be to deploy to a point at a set distance from home base, a force of a definitive size and character.

An example of a "Capabilities-Driven" paradigm would be ability to inflict "shock and awe" on any potential adversary. The shock and awe concept proposes to focus U.S. military strategy and war fighting capabilities on the ability to disorient both physically and mentally an adversary through the exploitation of advanced technology and new operational concepts. In theory, physical destruction and casualties could be minimized by the impact of shock. The awe inspired by such a capability would be the basis for deterrence and dissuasion.

A capabilities-based approach to force planning would lend itself to a resolution of long-standing roles and missions issues. The future force would be designed to meet specified requirements in capability categories (e.g., control of airspace, deep strikes, battlefield interdiction, close fire support, etc.). Competing capabilities could be evaluated against one another for that mission.

Concluding Thoughts.

Undoubtedly, there are innumerable, clever ways to recast the current two MTW paradigm to something that is slightly smaller, not quite as old, and, hopefully, less expensive. None of these constitutes a new paradigm. They allow the Bush administration to continue on the current path in force shaping. A different paradigm must start with a different strategic perspective on the nation's national security objectives and on the role of military power in meeting them. Anything else amounts to rearranging the deck chairs on the Titanic.

ENDNOTES - CHAPTER 6

1. William S. Cohen, Secretary of Defense, *The Report of the Quadrennial Defense Review,* Washington, DC: Department of Defense, May 1997, p. 8.

2. Report of the National Defense Panel, *Transforming Defense: National Security in the 21st Century*, Washington, DC: Department of Defense, December 1997, p. 23.

3. Mackubin Thomas Owens, "The Price of the Pax Americana," *The Wall Street Journal*, October 16, 2000, p. 26.

CHAPTER 7

WHY THE 2 MTW MUST GO[1]

Michael Casey

At the beginning of the 21st century, the United States finds itself standing, almost by accident, in a preeminent position in a world wracked with uncertainty, where the only assurance is that the global situation will continue to change. America faces both a great opportunity and a great danger: the opportunity that the world could be "shaped" or encouraged to become more democratic and peaceful, or if efforts to encourage such fail, the danger of a world filled with conflict, existentially-threatening transnational threats, and possibly the rise of a country or countries which desire to challenge, eliminate, or replace the United States. In the hopes of encouraging the opportunity while avoiding the danger, the United States has adopted what can best be described as a hedge military strategy, usually referred to as "Shape, Respond, Prepare." This strategy has, at its base, the intention of providing stability throughout the world today while preparing for worst-case scenarios in the future.

There is a fundamental internal conflict to this strategy which must constantly be reevaluated and balanced. That is, how much emphasis should be focused on each of the various elements? Currently the U.S. defense establishment feels that it must maintain sufficient forces to fight and win two near-simultaneous major theater wars (or two major regional contingencies). This chapter will argue that this, referred to as the "2 MRC" force-sizing paradigm, is an inappropriate model for the current U.S. strategy as it is badly linked to real world threats, is largely unaffordable, and short changes the future.

With the collapse of the Soviet threat to Europe, decades of force planning constructs were suddenly obsolete. Casting about for a replacement by which to measure needed military forces, the Bush administration and Congress, as described by Eliot Cohen,[2] reduced military forces by 30 percent while maintaining a high operations tempo and research and development budgets, but while reducing procurement rates. Intended as cautious policy to tide the nation over until it was fully clear that the Soviet threat was truly gone, the Base Force was a rational hedge. Following the Gulf War, Secretary Les Aspin of the new Clinton administration formalized Pentagon thinking by evolving the capabilities-based force of the Base Force into one capable of facing two, relatively well-defined threats in near-simultaneous major regional conflicts. It was envisioned that these conflicts would require roughly the same forces as the just-fought Gulf War, only doubled.[3] Several years later, the 1997 Quadrennial Defense Review (QDR) accepted this two MRC model as the current force sizing criteria.

It should be no surprise that the two MRC paradigm has been heavily criticized for making insufficient changes to America's military. Eliot Cohen has commented that, "The two MTW strategy requires almost the same force structure as did the Base Force; a dispassionate observer would be hard pressed to see much difference between the two."[4] Retired General Anthony Zinni, USMC referred to the current force structure as, "a mini-version of the Cold War force." Going further, he said, "Let's admit it—we've screwed up again."[5] Even Congress has gotten in on the act as stated by the House Appropriations Committee, "It is now all too apparent that the military services are not yet properly reconfigured from their old 'Cold War' orientation."[6] From these criticisms, if nothing else, it is readily apparent that whatever the merits of the intentions behind the evolution to the two MRC force, the changes simply have not gone far enough.

It may well have been true, as Department of Defense (DoD) planners feared at the beginning of the decade, that the United States could potentially face two countries who simultaneously attempted to invade their neighbors with classic combined arms, armor heavy forces. Certainly, in the wake of the Gulf War when the doctrine was first publicly outlined and the North Koreans were busy rattling sabers, this was a potential and real fear. However, in the year 2000, and for the likely foreseeable future, this scenario has become less likely. U.S. forces have become more advanced and capable, although stretched and overworked, while the adversaries against whom the 2 MRC has been judged have become increasingly weak and, in at least one case and possibly two, less politically motivated.

Iraq's forces have never been truly rebuilt after the Gulf War, the embargo has had a severely detrimental effect on the attempts which they have made to build stocks of war materiel, and some observers have even commented that the Iraqi military is roughly half the size it was prior to 1990. North Korea for its part has suffered the loss of its prime military backer, the Soviet Union, at least a decrease in overt support from China, a devastating famine, and severe economic hardship. Its military, while largely opaque to the outside world, likely has been affected by this, even if not as severely as civilian North Koreans. Additionally, the South Korean forces have, for their part, increasingly become more capable and technologically advanced. North Korean leaders have even begun making noises about rapprochement with South Korea and the United States. While it is possible that this is simply a diplomatic ploy, if real it is not the kind of move a country with overwhelming military superiority typically makes. A third country, Iran, sometimes mentioned as a potential MRC, has plenty of its own problems. Its military was devastated by the Iran-Iraq war and has never really rebuilt. It has a veritable cornucopia of old U.S. and Russian military equipment that has reached the point of obsolescence where it would serve adequately only in a museum. Additionally, its revolutionary Islamist government has been defeated in election after election, its reformer president has made diplomatic outreaches to the West, and it has its own economic problems. While certainly still capable of doing damage to neighbors or oil shipping, Iran looks to lack the military capability to seriously attack anyone but Iraq.

All three of the countries mentioned above have, it is certainly true and will be brought up later, substantially increased their capability to carry out asymmetric attacks, but not one of them is truly as great a threat for a cross border invasion as they were 10 years ago. These facts have not gone unnoticed by military and political commentators. John Troxell in *Force Planning in an Era of Uncertainty* comments, "The Iraq and North Korean scenarios remain the most demanding, but in each case, threat capability is declining."[7] Secretary Cohen himself said, "Saddam Hussein is probably at half the strength he was prior to the Persian Gulf war."[8] Perhaps summing up the disparity between U.S. conventional forces and our potential adversaries, Jeffrey Record says, "How dumb would an enemy have to be to take on the United States at the conventional level of warfare?"[9]

The 2 MRC assumptions, beyond the basic one that two enemies might simultaneously attempt to commit military suicide by attacking U.S. allies with a classic combined arms force, are also certainly unrealistic. First, the 2 MRC defines both scenarios as requiring an approximately DESERT STORM sized force—five divisions or so, several air wings, three to six carriers, etc. Given the relative decline in potential adversaries' military strength and increases in U.S. allies military strength, not to mention U.S. military capability, this is unrealistic on the face of it. Secondly, for some reason planners have chosen the cross border invasion scenario with combined arms as the most stressing, even though it is easily apparent that this is not so. Far more stressing, for example, would be a North Korean chemical and biological attack delivered with missiles. Not only would this be devastating, but if ports and airfields were targeted too, U.S. forces would be unlikely to even be able to make it in the theater quickly without substantial losses. North Korea's missile capability is, after all, at least the public threat for which the United States requires a national missile defense. An attack on Taiwan by China could be far more stressing for the U.S. Navy than any Operation DESERT STORM scenario. Even a Kosovo II, if repeated using air-only forces, would likely stress the U.S. Air Force more than "Son of Desert Storm." The United States has, after all, been flying over and bombing Iraq for 10 years without a loss, and it is unlikely that doing so en masse would be harder than a war from unprepared bases a long way from the front against a dispersed, light force using innovative tactics. Third, the 2 MRC force has, as too many commentators to list have noted, not provided the organization and forces the United States requires for the most common mission which the military undertakes, that of the SSC—Small-Scale Contingency, sometimes referred to as "nation-building." Fourth, the 2 MRC force construct has not provided the United States with those forces which can truly defend it and its allies against what are referred to as "asymmetric attacks": terrorism, WMD, missile strikes, computer warfare, etc. Until the cross border invasion focus is removed, it is unlikely to do so.

Maintaining the 2 MRC force and keeping the required forces at a high level of readiness, even without addressing the future threats mentioned above and dealt with more below, is expensive and will likely require relatively large investments in the future. The Congressional Budget Office has concluded that merely sustaining the current force would cost about $50 billion more than the United States currently spends on defense.[10] Various other commentators have concluded that it would require between $25 billion and $100 billion more per year. Commandant of the Marine Corps General Jones has proposed that annual defense spending rise to above 4 percent of Gross Domestic Product per year, about a $100 billion increase.[11] Air Force Secretary Whit Peters has also been quoted as calling for increases of $100 billion to the military budget.[12]

Potentially worse than the amounts required to simply maintain the 2 MRC force, is the opportunity cost presently of that force. There seems to be little doubt that the need to keep the force up to par to be able to "respond" is squeezing out the investment needed to "prepare now." Time after time, planned increases in modernization accounts have been pushed back due to shortages in readiness accounts. Pay increases and military health care accounts have attracted the healthiest share of recent defense budget increases. As General Jones puts it, "We are, in essence, continuing to maintain our current status at the expense of future readiness."[13] The future readiness to which General Jones was referring will only come about if the U.S.military is modernized, to include procuring new platforms and systems to replace those currently wearing out.

There is relatively little dispute, in addition, that the United States does, in fact, need to "prepare now." In the same CBO publication which outlined the $50 billion shortfall in defense spending, the authors note that, "As military leaders have noted, however, the most worrisome threats to U.S. interests may not be the conventional forces of foreign powers. Instead, so-called unconventional threats to the United States and to U.S. forces may pose greater dangers." They further went on to note that, "Many of the same regional powers that the United States planners worry about facing in conflicts with conventional weapons also have NBC [nuclear, biological, chemical] weapons."[14] The National Defense Panel has noted, ". . . our current course is unlikely to produce the military capabilities necessary to meet the range of challenges we foresee in 2010-2020."[15] The Center for Strategic and Budgetary Assessments, as well as the Pentagon's own Office of Net Assessment, have both issued literally more statements on this subject than can be mentioned. Suffice it to say, almost no observer of security affairs believes that the threats of

2010 or 2020 will be those which are faced today or were faced when the 2 MRC force shaping paradigm was adopted. And, as an almost equal number of observers have noted, current readiness demands are squeezing out the efforts to prepare for these future threats.

The lack of ability to keep up current forces while addressing shortages in the capabilities presented by the two MRC force could probably be easily fixed if the United States could resort to its traditional tactic of throwing money at a problem until it goes away. Unfortunately, this is probably not possible, and is certainly unlikely. The much ballyhooed surplus is, at current rates of spending and desired tax cuts, unlikely to ever materialize on anything approaching the scale mentioned in press reports and seemingly accepted throughout Washington. For example, the Center for Budget and Policy Priorities has concluded only about $400 billion of the surplus will likely be available for new tax and spending programs, not $1.9 trillion.[16] The Brookings Institution, calculating the available surplus slightly differently, concludes that only $350 billion will be available.[17] And both of these estimates do not take into account popular initiatives such as reducing the marriage tax penalty, a Medicare drug plan, education reform, or pension reform, which would likely come before substantial increases to the military budget were considered.

It is also interesting to note that neither presidential candidate proposed increasing military spending beyond about $10 billion per year. John Kreul commented, ". . . both presidential candidates already have committed most of the budget surplus to other areas. . ."[18] Similarly, Roberto Suro said, "Boosting the Pentagon budget by that much [$50 billion a year] would require both candidates to rethink domestic programs."[19] Congress too, is unlikely to be enthusiastic about huge military budget increases that threaten domestic priorities and tax cuts. The current budget resolution shows the 050 Defense account getting $307.3 billion in FY01 and $332.5 billion in FY05.[20] After inflation, this is not a huge increase, even accounting for the reality that the FY01 budget resolution has been violated repeatedly. Over the past few years, Congress has not added huge amounts to the defense budget despite repeated pleas from military commanders, and that which has been added has frequently gone to what many observers regard as "pork." Kori Schake noted,

> . . . the $30 billion that have been added since fiscal year 1996 to the defense budget are not predominantly helpful to our national interest or to carrying out our defense policy. Two billion of that alone is military construction when we need to be shutting unused bases, not adding on.[21]

Nor are huge increases likely in the future. John Lumpkin quotes Senator Pete Domenici, Chairman of the Senate Budget Committee, referring to the CBO study which concluded that $50 billion per year was required to maintain current capabilities, "It doesn't necessarily mean the military should simply get the extra money," he added. "Budget gaps such as this have been a reality since the 1970s," he said.[22]

The force in existence under the 2 MRC policy is not developing the capabilities the United States will need to meet future threats, is expensive and unlikely to receive much fiscal relief, and is designed around an uncertain and increasingly unrealistic threat. So, what should replace it? The answers are, unfortunately, not as clear cut as the relatively simplistic 2 MRC paradigm and will require substantial willingness to adapt to changing situations in the future. First, DoD needs to substantially expand the planning scenarios, particularly with an eye towards the "shape" and "respond" strategy elements and adapt the current force to better, or at least more realistically, deal with them. There needs to be an admission that these will not all require cookie-cutter copies of the DESERT STORM force. In addition to the Iraq and North Korea threats, DoD should look, for example, at a Chinese invasion of Taiwan, at an Iranian attempt to shut down shipping in the Persian Gulf, at potential conflicts in the Balkans, at regional wars in the Caucasus regions, a breakdown, including low level war and sniping at shipping in Indonesia, etc. All of these would require not only different forces than the current scenarios, but may well require realignment of current basing for the shaping function.

For example, in the Iraq scenario, the United States currently keeps an armored brigade in Kuwait 9 months or so out of the year, as well as composite air wings in Saudi Arabia and Turkey. It may be beneficial, to deter Iraq, to keep an armored brigade in theater year round, particularly one shaped for independent action if necessary. It might also be beneficial to look at the possibility of stationing a similar brigade in Eastern Turkey (which would also present the opportunity to have some kind of land force able to respond in the Caucasus if necessary), at least part of the year. This, combined with the air assets in theater, would offer a substantial deterrent capability, particularly if combined with a missile heavy component of some kind, probably naval based. The Navy needs to admit that there is relatively little role for carrier forces in a respond situation in this region. During the Gulf War, three carriers were present, none got closer than 300 miles or so to shore, and naval air forces, for all their contributions, heavily relied on Air Force refueling. Additionally, the traditional role of a carrier, responding to a threat where the United States has no assets or cannot build up assets quickly, is moot in this context—a substantial land and air force is already in theater. It would be better to save a carrier for some other role and to supplement U.S. efforts in the region with missile shooters.

Similarly, in Korea, if war breaks out there is little likelihood of a fast reinforcement of U.S. forces there excepting carrier-based air power and naval missile attacks. North Korea would simply have to be blind or insane to not take steps to stop the United States from rapidly reinforcing South Korea. Their likely strategy would be to try to end the war rapidly, and U.S. forces should be aligned to respond to this. The two Army brigades in Korea would probably have to be able to operate on their own for some time and should be reconfigured for independent action. These would present theater commanders with an offensive capability and would absorb considerable North Korean attention and resources, while presenting a substantial deterrent. Similarly, it may be beneficial to keep more carriers in the region—the one freed up from duty in the Persian Gulf would be available. Repositioning a carrier to the Pacific to enable more coverage would also allow for this force to be used in a Chinese invasion of Taiwan if it was deemed necessary by Washington.

Second, DoD needs to take a serious look at how its organizational structures are stressing the force. For years, the Navy and Marine Corps have had a rotational force capable of keeping a substantial number of ships at sea without breaking the remaining force. The Air Force has recently moved to the Air Expeditionary Force (AEF) to try to do much the same thing. Only the Army does not plan to change to allow for forces designed for rotational duty. This decision should be reexamined. Douglas MacGregor's now famous book, *Breaking the Phalanx*, proposed moving towards smaller, brigade-sized units. The Army has chosen to not do this, but it is a decision that should likely be revisited. Divisions may be fine for fighting Iraq in the desert or the Soviet Union in Germany, but sticking with the divisional structure has led to a high rate of deployment for headquarters units in SSCs as well as high demand/low density units like civil affairs. A combined arms brigade type structure could solve many of these problems by essentially creating many more units capable of conducting a Kosovo or Bosnia type mission without augmentation. This would reduce the personnel tempo felt by many soldiers. This is the kind of analysis that DoD and the services need to be exploring, even if it eventually does not pan out.

Third, DoD needs to look to the kind of forces, capabilities, and efforts needed for the future and truly serious threats. Professor Ashton Carter of Harvard discusses what he calls "A-list" threats. These are those of Cold War scale, those which threaten the existence of the United States. None of the threats on Professor Carter's list, a Russian collapse similar to Weimar Germany, Russia losing control over its arsenal of weapons of mass destruction, a truly hostile China, massive proliferation of weapons of mass destruction among many nations, and catastrophic terrorism on U.S. territory,[23] are traditional conventional military threats save possibly the hostile China one. Yet the sheer scope and potential damage of any one of them demands that DoD pay much closer attention to them, and, where appropriate, augment capabilities or forces that can preclude such threats or, in the worst case scenarios, help mitigate them. It should be noted that, for all the potential damage of a Iraqi takeover of Kuwait or a North Korean conquest of South Korea, none of the current threat scenarios for which the U.S. military is prepared would come anywhere close to the damage caused by any one of Professor Carter's nightmares. Given this, it would seem to make sense that if risk has to be taken, it be taken more among the "B-list" threats than the "A-list."

Fourth, DoD must engage in a relatively large effort of experimentation. While any number of observers have commented on the need for transformation, it is not at all clear what precisely the U.S. military should be transforming into. There are threats and strategies which the U.S. military must, in the future, be able to combat. However, as another, smaller, group of writers have mentioned, it is vastly important that the United States get transformation right, rather than simply transform for its own sake. It would, after all, do little good to transform into the modern equivalent of the 1939 French Army. This argues for a substantial effort of experimentation. DoD, through the new Joint Forces Command and service-level efforts, should be encouraged, or if necessary forced, to try new platforms, systems, organizations, tactics, and doctrine. Most of these efforts will not work. All of them will teach some lesson and will be valuable for that. As part of an ongoing experimentation effort, DoD needs to avoid purchasing "lock-in systems," those which, due to high cost, large numbers, and/or long life spans, would essentially lock the United States into one organization structure, doctrine, or platform. While some replacement is vital due to aged equipment, a mid-term strategy for the rest might be to upgrade or extend service life where possible, rather than replace wholesale. Similarly, an ongoing, high-intensity effort in data links and communications would likely pay high dividends in the future by enabling more "leveraging" of various assets, sensors, etc. As part of the experimentation, DoD will have to acquire small batches of some systems with which to experiment and "play." An example might be the converted Trident "SSGNs" sometimes mentioned, which would give a short-term military capability as well as serve as surrogates for future systems in experimentation efforts. Given the long-term budget crunch facing DoD, it is also going to be necessary for various programs to compete with one another. For example, some outside DoD have suggested that the Joint Special Forces (JSF) program be "shifted to the right" or delayed several years. This could be combined with experimentation efforts in Unmanned Combat Aerial Vehicles (UCAVs) and the systems, possibly along with other air power platforms such as long-range bombers, could be placed in competition to see which has the best long-term benefit for the United States. DoD is highly unlikely to be able to afford all of its planned programs and something will have to give. Competition within experimentation seems likely to result in increased capability to some extent while reducing procurement costs in the long term.

Finally, given the budget forecasts, DoD is going to have to resize the force, if only to save money. However, if combined with the experimentation efforts listed above, this can also provide an opportunity to kick loose new ideas within the military. In the Navy, for example, carriers are in some cases, as briefly mentioned above, declining in utility. Yet the Navy continues to keep carriers on station in the Mediterranean, the Persian Gulf, and the Pacific. Of the three, only the Pacific makes sense. Both the Mediterranean Sea and the Persian Gulf host substantial U.S. Air Force and Army forces. Shifting some of these carriers to the Pacific Ocean for better coverage there while retiring two or so without replacing them would free up substantial resources for the pursuit of other systems which may be better suited for the future, such as the SSGN concept or the "Street Fighter" idea. The Army National Guard will also have to be rationalized. Currently, the Guard maintains eight divisions which have no war time mission. Retiring two of these would allow the resulting dollars to be put into increasing the readiness of the other six, which would allow them to serve as heavy reinforcements in time of war, particularly if the active duty Army trends more towards "medium weight" formations and/or experiments with brigade-based forces. It might be worth examining why the Marines maintain 20,000 men in Okinawa, but only have three transport ships there. One of those numbers has to give; the occupation of Japan is over. While it is useful to station forces there, it is only useful to do so if they can respond to a threat, which is unlikely to materialize on an outlying Japanese island. Either the number of transport ships should be increased, or, more likely, some or all of the Marines should be pulled out of Okinawa and put back into the fleet. Combined with a the likely slight increase, $10 billion or so, in the DoD budget and necessary efforts to close unneeded bases and see efficiencies in DoD's support organizations, such resizing of military forces may free up enough money to alleviate the budget crunch while simultaneously offering the opportunity to experiment to develop forces better suited for future threats.

In summation, DoD can no longer afford to maintain a force sizing model which short changes the future, is unaffordable, and is not clearly aligned with current efforts to either shape or respond to current threats. The upcoming Quadrennial Defense Review will offer a clear opportunity to eliminate the 2 MRC model and to move to one based on a more diverse threat universe, a realistic budget environment, and which provides opportunities to experiment to come up with solutions for future threats. While the 2 MRC may have served a useful purpose, that is over, and it is time to balance all the elements of U.S. national military strategy—"Shape, Respond, Prepare"—not emphasize one, "Respond" at the expense of the other two.

ENDNOTES - CHAPTER 7

1. This chapter is the result of the author's own thought process and should not be taken to be representative of the views of his employer, other Congressional staff, or the Congress as a whole.

2. Eliot A. Cohen, "Defending America in the Twenty-first Century," *Foreign Affairs*, November/December 2000, p. 42.

3. For more detail, see John. F. Troxell, *Force Planning in an Era of Uncertainty: Two MRCs as a Force Sizing Framework*, Carlisle, PA: U.S. Army War College, Strategic Studies Institute, September 15, 1997, especially pp. 9-13.

4. Cohen, p. 43.

5. General Anthony A. Zinni, "A Commander's Reflections," March 2000, reprinted at *http://www.usni.org/Proceedings/Articles00/prozinni.htm*.

6. House Appropriations Committee, *Department of Defense Appropriations Bill, 2000: Report of the Committee on Appropriations*, H.Rpt. 106-244, Washington, DC: U.S. Government Printing Office, 1999, p. 11.

7. Troxell, p. 30.

8. Testimony before the House National Security Committee, May 21, 1997. House National Security Committee, *The Quadrennial Defense Review*, H.N.S.C. No. 105-14, Washington, DC: U.S. Government Printing Office, 1997, p. 70.

9. Jeffrey Record, "America's Hollow Statecraft: Sources and Consequences," *Strategic Review*, Fall 2000, p. 50.

10. Congressional Budget Office, *Budgeting for Defense: Maintaining Today's Forces*, Washington, DC: Congressional Budget Office, September 2000, p. XI.

11. From Michael O'Hanlon, "How Much Money is Enough for Defense?" *Los Angeles Times*, September 25, 2000. Found on Internet at *http://ebird.dtic.mil/Sep2000/e20000926defense.htm*.

12. Reuters News Service, "Official: Military Budget Needs Extra $100 Billion," *Washington Times*, October 27, 2000, p. 13.

13. General James L. Jones, Written Testimony before the House Armed Services Committee, September 27, 2000, p. 5.

14. Congressional Budget Office, p. 3.

15. National Defense Panel, *Transforming Defense: National Security in the 21st Century*, Washington, DC: Department of Defense, 1997, p. 21.

16. James Horney and Robert Greenstein, "How Much of the Enlarged Surplus is Available for Tax and Program Initiatives?," Washington, DC: Center on Budget and Policy Priorities, July 7, 2000, p. 1.

17. Alan J. Auerbach and William G. Gale, "How Big Is the Prospective Budget Surplus?," Washington, DC: The Brookings Institution, September 2000, p. 1.

18. John Kreul, "Son of QDR: Prospects for the Army," *National Security Watch*, Arlington, VA: Institute of Land Warfare, Association of the United States Army, July 15, 2000, p. 3.

19. Roberto Suro, "Shelton To Present Military Choice," *The Washington Post*, September 27, 2000, p. 21.

20. Committee on the Budget, *Concurrent Resolution on the Budget—Fiscal Year 2001*, H.Rpt. 106-530, Washington, DC: U.S. Government Printing Office, 2000, p. 34.

21. Kori Schake, "Balancing Commitments and Capabilities," in *After Kosovo: Papers from a Capitol Hill Seminar*, Lucy Webster, ed., New York, NY: Economists Allied for Arms Reduction, 1999, p. 32.

22. John J. Lumpkin, "Military Doctrine Tackled," *Albuquerque Journal*, October 25, 2000, p. 1.

23. Ashton B. Carter, Written Testimony, "The Perils of Complacency: Adapting U.S. Defense to Future Needs," submitted to the Subcommittee on Emerging Threats and Capabilities of the United States Senate, March 21, 2000, p. 4.

CHAPTER 8

MULTI-METRIC FORCE SIZING

Ian Roxborough

America's armed forces are tasked to fight and win the nation's wars. Warfighting is their core competency. There is, moreover, a clear vision of what the current American way of war entails. It is embodied in *Joint Vision 2010* and supporting documents, and in the canonical scenarios of halting cross-border aggression by armored forces. This focus on the core competency provides a clear sense of purpose and a clear metric for force sizing exercises: the two-major theater war (MTW) standard. Unfortunately this tends to produce a one-size-fits-all approach which is inappropriate for the current strategic situation. This chapter argues that the United States should identify the basic tasks required of its armed forces in the current strategic situation and design forces accordingly. The result should be a set of organizations with differing core competencies, and different metrics for force sizing.

Current U.S. defense planning treats most military forces as all-purpose forces (capable of executing a wide range of missions) but basically designs those military forces to fight a specific kind of war (cross-border armored aggression.) Since the tasks of the military are broader than warfighting, and since wars vary greatly in their nature, this type approach has serious drawbacks. (Some U.S. military forces are, of course, tailor-made for particular kinds of operations. Special Forces and nuclear deterrent forces are in this category. But most U.S. military forces are designed as general purpose forces.) This chapter recommends considerable increase in "tailoring" of U.S. forces in order to meet the range of strategic tasks which now confront the nation. Task-tailored forces (with several core competencies) rather than all-purpose forces (with war as the core competency) should be the prism through which force-sizing exercises operate.

Current Arguments for a 2 MTW Force.

Several arguments have been put forward to justify a 2 MTW force. The most well-known is the argument about the need to deter opportunism—"avoiding a situation in which a potential regional aggressor might be tempted to take advantage when U.S. forces are heavily committed elsewhere."[1] In principle, this is not unreasonable. But deterrence of opportunism requires only that the potential regional aggressor be persuaded that he will not gain from aggression. For this purpose the United States needs either the capability to slow down the aggression in the second region until it can shift forces from the first theater or it must convince the second potential aggressor that any gains that it might make will surely be rolled back at some point in time when the United States chooses. In either case, the metric for sizing forces for the second MTW would be different, and probably smaller, than in the first MTW. Given the small number of potential regional aggressors in the world, the probability that a second aggressor will be capable of, prepared for, and motivated to engage in cross-border attack at the same time the United States is heavily engaged in the first MTW is surely not very high.

A second argument for a 2 MTW force is that it "helps ensure that the United States will have sufficient military capabilities to defend against a coalition of hostile powers or a larger, more capable adversary than is foreseen today."[2] In other words, being able to fight two major wars will help the United States to be able to fight one really large war. The truth of this is indisputable, but a careful assessment of this argument hinges on the word "help." Of course larger forces would help prepare the United States for a really large war. But simply to argue along the lines that more is better avoids the question of whether a really large war might have different force requirements than two MTWs. If a really large war were to occur in Asia, for example, the United States might want a greater proportion of its forces with longer reach, with more transportation assets, and with greater capabilities for replacement and reconstitution. It is not necessarily the case that forces designed for two MTWs are the best forces for one really large war. Undoubtedly there are trade-offs here, and these should be addressed explicitly.

A third argument for the 2 MTW standard is reassurance of allies and general promotion of U.S. interests. There are many ways that the United States can reassure allies and promote its interests. How do we know whether an increment in military forces does this better than the same amount of money spent on some other activity, such as more diplomacy, payment of U.N. dues, or something else? This argument is a hang-over from the Cold War days when there was a large U.S. overseas presence. The forces stationed forward to meet the Communist threat could also be seen as promoters of more general U.S. interests. Given the forward presence of considerable U.S. forces, the marginal cost of using these forces for the promotion of a wide range of national interests was minimal. As this presence continues to decline, the means by which allies are reassured and general U.S. interests promoted will need to be reassessed. It is now too expensive to undertake these tasks largely with military forces.

A fourth argument—sometimes conflated with the third—is that, as a global power the United States needs a force that demonstrates its status as a superpower. This is a reputational argument. As such it tends to imply that because reputation was enhanced in the past because of the size of America's armed forces, in the future size will continue to be the best way to maintain this reputation. The argument about reputation is inherently conservative. It suggests that the United States should continue to do things the way it has done in the past because any change might be misunderstood and lead to a loss of reputation. American strategists need to resist this temptation, and need to ask what other ways America's reputation as a great power can be maintained.

Strategic Context of Force Planning.

Decisions on force sizing follow logically from a prior assessment of the strategic situation. As we know, in the years since the end of the Cold War this has been a widely debated issue, and no consensus has been reached. A very wide range of existing and potential threats have been identified. There is no agreement over the relative importance of such threats. The result, in terms of force sizing, has been two-fold: on the one hand the United States has experienced a great deal of inertia in restructuring its forces. On the other, such innovation as has occurred has taken the form of a series of ad hoc and poorly assimilated responses, particularly in organizational terms. There have been efforts at embracing the current revolution in military affairs (RMA), at improving jointness, and at driving transformation of existing organizations. These efforts have often been moves in the right direction. However, some critics feel that these transformation efforts have failed to achieve the synergy and dynamism that is needed.

The forces that were in place at the end of the Cold War have been reduced in size, but not fundamentally changed in structure. They are smaller and more up-to-date versions of the Cold War force. The trend has been evolutionary. And the canonical scenarios around which post-Cold War force planning has revolved—the 2 MTW force based on the illustrative cases of Iraqi and North Korean cross-border attacks—are simply scaled-down versions of the Cold War planning assumption of a massive Soviet attack through the Fulda Gap. The great success in the Gulf War reinforced this tendency to think of war in this way. Much current U.S. force planning continues to revolve around the image of halting and reversing an armored cross-border attack. The strategic appraisal implied by this force structure is that of a series of regional threats plus a motley list of other threats to worry about. Implicitly, the items other than defeating cross-border aggression on the constantly changing list of threats are usually treated as lesser-included cases which can be managed by the forces designed to deal with major theater wars.

There is a clear disconnect between the identification of a wide range of diverse threats, and the force-sizing activities of the Department of Defense (DoD). Force-sizing has focused primarily on the regional threat issue; other kinds of threats, while duly noted (and sometimes highlighted) in the official reports, seem to play only minor roles in force-sizing exercises. Such issues as homeland defense and peace operations have considerable implications for the types of forces the United States needs, and it seems reasonable to assume that forces appropriate to these missions will look rather different than forces designed to fight major theater wars. Basing force-sizing exercises primarily on the need to fight major theater wars cannot be a sound way to decide what forces the United States needs. It is a case of the tail wagging the dog: because force-sizing on the canonical scenarios of Korea and Iraq is so comfortable, so similar to Cold War assumptions, the regional threat has driven force planning to the virtual exclusion of other threats and needs. Regional threats are but one factor among several that need to be considered in designing U.S. military forces. The full range of threats that have been identified needs to be considered, with some explicit discussion of relative priorities. In addition, there is a need for the employment of military forces which exists independently of "threats." Many peace operations, and overseas presence, are only a response to "threats" if the term is stretched so far as to lose any real meaning. Strategic assessment cannot simply be threat-based.

Although this chapter argues that the metric for force sizing should derive from a strategic assessment, space considerations preclude an extended discussion of strategy. Nevertheless, the strategic assumptions underlying this chapter should be laid out explicitly. They are: (1) we are currently in a strategic pause; (2) there is a wide range of diverse threats; (3) there is a need for a range of military forces to engage in activities only indirectly related to specific threats; (4) there is a need to prepare for a war of greater dimensions than the current notion of an MTW, if only as a deterrent measure.

The Case for Experimentation.

The implication of a strategic pause is that now is the time to experiment with doctrine and force structure. Of course, from the perspective of policymakers and commanders in chief (CINCs), U.S. forces are constantly responding to one crisis after another. Stretched thin, unable to accomplish pressing missions adequately, the notion of strategic pause appears like a figment of the imagination to the hard-pressed people in the thick of the action. To overcome the pressures of the present, we need to design organizational forms that enable commanders and strategists to focus on the long term.

Experimentation, rather than acquisition or force structure, ought to be the top priority for defense planners. Experimentation should not be equated simply with furthering the RMA. The new American Way of War as embodied in *Joint Vision 2010* and its successor documents is fundamentally a preference for high-tech war. While this is clearly the right approach overall, the U.S. military needs to experiment with a lot of low-tech operations as well. And experimentation should focus as much on organizational design and warfighting concepts as on tactical issues.

There are two important arguments why experimentation should be a high priority. The first is the likelihood of "getting it wrong." When addressing innovation in a context of rapid technological change, the natural trend is to expend effort in getting it right. Because the future is so unpredictable, and because enemies will be working equally hard to thwart U.S. innovations, the likelihood is that whatever innovation the United States designs now will turn out not to be quite right when the time comes to employ it in combat. The risk of not "getting it right" is high, and the only way to guard against the negative consequences that might follow is to create a military which is adaptable and flexible, ready to take failure in its stride. The lesson of America's first battles is not (contrary to much current thinking) that the United States should work harder to prevent failure in the initial phases of a campaign, but rather that the U.S. military needs to take a hard look at how it can respond to initial failure and go on to adapt for ultimate success. It is more important to be adaptable than to be right the first time.

The second argument is a logical corollary of this, and has to do with the need to nourish and sustain a substantial number of flexible thinkers in the military. The most valuable resource in a time of rapid technological change is a pool of innovative thinkers. Force design—particularly with regard to retention and reserve issues—should explicitly aim to foster the existence of a large body of people with the right attitudes. Many of the skills are embedded knowledge, learned from years of experimenting with new concepts. They will exist at all levels of the military, not just among the ranks of those who have attended the War Colleges. Many sergeants and technicians will have important aptitudes and skills which will be critical for rapid innovation and recovery from initial mistakes. Experimental forces should be nourished and the personnel in them should be carefully tracked to maintain a core of experienced people who can pass on their ingrained knowledge to others.

Dealing with a Wide Range of Threats and Tasks.

As numerous analysts have argued, the end of the Cold War has meant a shift from one dominant threat to a multitude of existing and potential threats, and it has increased the salience and frequency of some nonthreat related military missions such as peace and humanitarian operations. The Department of Defense has had difficulty defining a list of such threats: at times it seems as if it had produced a shopping list, with items added or deleted without any underlying rationale being easily discernable. Part of the problem is that status quo powers, such as the United States today, tend to regard any change as "instability," and see instability as a "threat." This temptation should be resisted.

Expansive definitions of national security have emerged in recent years. It has been argued, for example, that drugs are a threat to national security. Drug trafficking may be undesirable, and the amount of drug usage by the U.S. population may be a cause of concern, but it is hard to see what is meant by defining this as a threat to national security. There is no need to argue that an issue is a national security issue in order to involve the military. Whether or not the military should be used should be decided entirely in terms of whether they or some civilian agency are better equipped for the task at hand. To argue that drug trafficking is a national security issue is to identify any issue that concerns the well-being of the American population with national security. If we accept this, there is nothing that is not a national security threat, and the term becomes meaningless. To give another example, in the early 1990s it was argued that a failure to maintain a healthy U.S. economy posed a risk to U.S. national security.[3] In one sense, this is absolutely correct. The links between the economy and defense strategy are many and intimate. A strong economy is a necessary precondition for an effective defense. But a necessary precondition of something is not the same as the thing itself. To put "a healthy economy" in a list with other threats such as regional aggression is to mix apples and oranges. We must not confuse what is a desirable state of affairs for the United States (and the world as a whole) with threats to U.S. national security. We teach, particularly at the War Colleges, that all elements of national power should be coordinated, that wars involve economies and societies as well as militaries. But while this is true, it does not follow logically that anything that makes the world a less desirable place to live in is a national security issue. What is lacking here is a reasoned argument about the long-term requirements for U.S. national security, and the relationship between changes in the global security environment and U.S. capabilities (including the health of the economy) to respond to such changes. Shopping lists seldom promote careful consideration of matters such as the relationships between long-term and immediate concerns, the relative importance of different issues, and trade-offs between responses to different concerns.

The United States has strategic goals which are not driven by immediate threats. They include such things as showing the flag, humanitarian assistance, maintenance of global peace, support of diplomacy, and promoting democracy, free markets, and human rights. These activities require substantial military force. Some of these tasks, such as showing the flag, can be done by forces designed to fight major theater wars. Naval overseas presence is a good example of this. With some other tasks, such as constabulary and nation-building operations, a case can be made that the forces designed for major theater war are not well-suited to these tasks. This chapter argues that tailored forces should be developed for such tasks, and that the warfighting components of the military should be allowed to concentrate more on their core competency.

One of the major complaints today is that America's armed forces are too busy. High PERSTEMPO is frequently cited as a reason for reduced morale and reenlistment. The costs of retraining units for warfighting after they return from peace operations are high. Tailored forces will be less expensive and more effective. There will be the added benefit of reducing the strain on the warfighting components of the armed forces. Tailored forces, of course, can only be justified if we foresee a long-term need for such forces. This is why strategic assessment is so central.

Larger War.

It need not take a peer or near-peer competitor for the United States to find itself engaged in a massive war that is much larger than the current MTW scenarios. Japan in 1941 was not a peer competitor of the United States by any stretch of the imagination, yet defeating Japan was a lengthy and costly affair. It is by no means beyond the bounds of possibility to imagine the United States becoming embroiled in a war against a powerful and resilient enemy (or a coalition) that demanded resources far beyond those needed for an MTW as currently envisaged. As I have already suggested, the forces required for a large war might be substantially different from those needed for two MTWs. Especially if the war were to occur in Asia, the problem of immense distance would require considerable logistical support. Anti-access and asymmetric strategies on the part of the adversary could well mean that the war became a prolonged one, with periods of relative inactivity and spikes of intense fighting. In such a war the United States would be at a disadvantage.

Current thinking, based on the "lessons" of the Gulf War, focuses on the need to get to the theater rapidly and win quickly. While these are appropriate concepts for an MTW, they may prove counterproductive when confronted with a well-prepared enemy that intends to fight a protracted war. In this scenario, the key issues for the United States will be sustainability and the rapid replacement of various platforms. Key assets like satellites, Airborne Warning and Control Systems (AWACS), joint surveillance, target attack radar system (JSTARS), and air refueling tankers may be targeted and may need to be replaced rapidly. Key munitions may be expended rapidly and stocks may be depleted. Replenishment of forces in the theater may be difficult. Current thinking about future warfighting tends to focus too much on rapid, decisive victory, and underemphasizes the possibility of protracted war. Strategists and force planners need to spend some time thinking about the kind of industrial base that will be required in such a war, the appropriate use of reserves, and how to move away from reliance on complex machines that take many years to produce. The United States might be better served by an ability to turn out large numbers of standardized machines rapidly, as was the case with the revolution in shipbuilding which produced the Liberty ships of the Second World War. Cheap barges filled with cruise missiles may be preferable to complex destroyers with elaborate defensive systems.

The United States might also be better served by a Reserve system that was able to rapidly integrate civilians into certain kinds of military operations (such as the myriad white-collar jobs that a computer-reliant military generates) without necessarily putting them into uniform or having them train on weekends. A "virtual Reserve" of computer-literate civilians might be a useful adjunct to the existing Reserve units. The experimental forces would be crucial cadres for force expansion.

Organizational Implications.

The argument of this chapter is that the United States should design a tailored force for each major strategic mission. What might this look like?

One obvious route forward is further development of the unified command plan. Joint Forces Command (JFCOM) is a great example of the kind of new command that could be formed. JFCOM is tasked with training and experimentation. It could also be tasked with preparing forces, reserves and the industrial base for a very large war. This sort of functional organization might be the way ahead. It is easier to make organization change through the unified command plan than through efforts to reform the services. Other new commands might be established, such as a Homeland Defense Command, an Expeditionary Command, and a Constabulary Command. Existing functional commands like Space Command and Special

Operations Command would round out the picture. Correlatively, the relationship between the geographical commands and the functional commands would need clarification. They perform different tasks, and it is by no means clear that they would be equal partners. The bulk of the forces would be assigned to the functional commands, and the geographic commands would be reduced to contingency and presence forces. Their main task would be to provide regional intelligence and expertise.

The bulk of the regular general-purpose warfighting forces which are currently configured to fight two MTWs would be assigned to the Expeditionary Command. These forces would be scaled to fight one MTW. JFCOM would continue as trainer and experimenter, with the additional task of preparing for a large war. The forces assigned to JFCOM would expand greatly.

The biggest change would be the creation of the Constabulary Command. This is perhaps one of the most difficult areas to reorganize. Peace operations need a wide range of skills, and the boundaries between military and other governmental action, and between government and nongovernmental action are usually blurred. The issue of international legitimation and the frequent need for coalitions of one kind or another raise many issues of command and control of military forces.

Rather than use regular U.S. Army troops for such missions, a dedicated force could be formed. It could be modeled on the Coast Guard: under civilian direction most of the time, but militarized in time of war. It would contain large numbers of aid workers, engineers, medics, etc., together with a strong core of area specialists. It would contain its own organic firepower, but largely of a light infantry and military police type. A good name for this unit would be "Peace Corps," were that name not already taken. Perhaps it might be called the "Peace Army." Unlike the existing Peace Corps, the constabulary force would consist of organizations rather than individuals, and members would be expected to serve for longer periods than is typically the case with the Peace Corps. To give an idea of what the Peace Army might look like, imagine a patrol in a troubled region where the Peace Army was deployed. A couple of light infantrymen and a Special Forces soldier would provide most of the firepower, in addition to their ability to call in air assets. The patrol would be led by a political officer, probably a Colonel, with long experience in the region. There would be an anthropologist, a doctor, two civilian police officers, and an nongovernment organization (NGO)-recruited liaison officer as the core components of the patrol. Most of the members of the patrol would be sergeants or officers. It would most definitely not look like a regular Army patrol.

It might be objected that recruiting and retaining personnel for this sort of mission would be very difficult. They would be seen as second-class soldiers. It is precisely the fact that they would not be doing "normal" warfighting tasks that underpins the argument for a tailored force. Peace operations are often very rewarding to soldiers who participate in them. By making it clear that they are doing a very different kind of task than warfighting—not an inferior task, just a different one—the role conflict between soldier/warrior and soldier/peacekeeper can be reduced, the costs of retraining can be minimized, the current complaints about perstempo can be addressed, and peace operations can be carried out more effectively with forces that are better suited to the task.

Another innovation would be the Homeland Defense Command. Defense of the United States is sufficiently important that it should not be seen as a subsidiary mission, or given to the Guard/Reserves to handle. Some aspects of homeland defense require a recognizable military organization as the appropriate response. Defense against air and missile attack is a good example. But other kinds of threat, such as cyberattack, or attacks by terrorists armed with weapons of mass destruction, require new forms of hybrid organizations. They require close intermeshing of military and civilian expertise. There is a need to examine carefully things like horizontal accession and short-service contracts as well as temporary secondment of civilians to the Homeland Defense Command. The blurring of boundaries between civil and military in a domestic context raises all sorts of thorny constitutional questions. Nevertheless, a clear case can be made for such a command, and hybrid civilian-military organizations may well be the wave of the future.

Although the organizational changes recommended in this chapter may seem radical, it should also be born in mind that the Department of Defense is evolving in precisely this direction. Force planners should simply be more explicit and more experimental about organizational design, and should attempt to introduce organizational innovation much more rapidly.

Some Downside Risks.

Change carries with it certain risks. But so does maintenance of the status quo in a changing environment. Keeping the two MTW metric carries risks. It only *seems* less risky because it is familiar. But if the strategic situation is changing, then comforting continuities are more dangerous than efforts at change. Some of the risks associated with the kind of approach to force-sizing that is advocated in this chapter can be foreseen. Others, of course, will come as a surprise. Let us briefly consider the more obvious foreseeable risks.

With regard to the emphasis on experimentation, the central risks are (1) being caught unprepared strategically and losing the first battle because forces ready for combat have been slimmed down too much, (2) inadequate force structure and acquisition, (3) run down of the defense industrial base, (4) loss of skilled labor force and of trained and ready soldiers, (5) the dreadnought effect of fielding weapons so advanced that all existing weaponry is obsolete and all militaries start from a level playing field, (6) complacency, (7) doing R&D for the rest of the world, and (8) lack of organizational focus at the level of component parts of the military: organizations need a clear mission and a single, simple measure of effectiveness.

With regard to the proposal for organizational change to produce more tailored forces, the principal foreseeable risks are: (1) lack of flexibility, (2) redundancy and underutilization of forces, (3) difficulty of integration of one set of tailored forces with another, differently tailored, force, (4) lack of unity of purpose and organizational focus at the level of DoD as a whole.

These are all risks. Some will exist if the current two MTW force is maintained. Some can be guarded against relatively easily. For example, the risk that a focus on experimentation will lead to insufficient acquisition to maintain a robust industrial base is not as serious as may seem at first sight. Experimentation will require some acquisition; it is not possible to experiment in a meaningful way with only one or two prototypes.[4] What experimentation will mean, however, is expensive unit costs because production runs will be small. Maintaining an expansible defense industrial base in these conditions will be expensive. Let us take another example: there is the risk that experiment in organizational design, together with the restructuring of the unified command plan proposed here, will lead to organizations without clear direction, pulling at cross-purposes with each other. True, this may happen. But this is already the existing reality. The current service-dominated organizations frequently pull at cross-purposes. Why would we assume that things would be worse with a different organizational design? They might well be better. We won't know until we try.

It might be said that a U.S. emphasis on experimentation will mean that America does research and development for the rest of the world. Well, this is already the case to a large extent. What is important is not simply the technical end result of experimentation, but the embodied knowledge and skills in the designers and operators. Experimentation produces smart, innovative people, and this is as much a resource as the tangible results of the experiments themselves. It will be vitally important to keep these cadres intact, replenish them with new people, and pass along the knowledge. This issue is not a new one: the need to maintain the embodied knowledge of a cadre of nuclear weapons designers, for example, has long been recognized. The proposal here is to extend this notion of husbanding embodied knowledge to a much wider range of military activities.

It might seem as if I am trying to solve the inevitable struggle over priorities and resource allocation by rewriting the organizational chart. Of course no rewriting of an organizational chart will abolish conflicts over priorities and resources. They are what strategy is largely about. There will be budgetary struggles between the new unified commands. The central problem is one of how the political system works to allocate resources among competing military demands. If the civilian political leadership cannot adjust budgets to align forces to changing strategic priorities, then whatever the organizational structure of the armed forces, there will be inevitable mismatches between strategy, budget and force structure. We know that this is a problem that exists today. It is largely a political problem. It can only be solved by reform of the budget and strategic process at the highest levels. But discussion of how America's political process might be improved to facilitate better strategy and force planning is beyond the scope of this chapter. The reorganization of America's military to produce tailored forces will not make the problem of how to redirect budgets to conform to strategic priorities (rather than to historically accepted divisions between the various services) go away. But to say that this chapter cannot propose a solution to all organizational and political problems facing the military does not mean that all organizational arrangements are equally effective. America's strategic situation is complex; it requires a range of measures of effectiveness, and several metrics for force sizing. The United States is now in a time of strategic pause; it can experiment with organizational structure in an effort to produce organizations tailored to the new tasks they will have to confront. It should do so.

ENDNOTES - CHAPTER 8

1. William S. Cohen, Secretary of Defense, *Annual Report to the President and Congress*, 1998, p. 8.

2. William S. Cohen, Secretary of Defense, *Annual Report to the President and Congress*, 1997, p. 6.

3. Les Aspin, Secretary of Defense, *Annual Report to the President and Congress*, 1994, p. 3.

4. Eliot Cohen, "Defending America in the Twenty-first Century," *Foreign Affairs*, November/December 2000, pp. 47-49.

CHAPTER 9

THREATS, STRATEGY, AND FORCE STRUCTURE: AN ALTERNATIVE PARADIGM FOR NATIONAL SECURITY IN THE 21st CENTURY

Robert David Steele

While suffering substantial reductions in manpower, and failing to modernize the conventional force, the American military claims to be ready so as to support the political claims of its current master in the White House. This claim does not stand up to scrutiny. The American military is not ready, either for two simultaneous theater conflicts, or for a range of Operations Other Than War (OOTW). In fact, we have real culture shock within our military, where a serving Chairman of the Joint Chiefs of Staff can be heard to say "Real men don't do OOTW" at the same time that units are stretched to the breaking point while they do exactly that: OOTW in every clime and place. The other elements of our national power—the diplomatic, economic, cultural, and justice elements of our government—are also not ready to make their contribution to national security in the 21st century.

We require a comprehensive evaluation of the threat, a reconstitution of our national security strategy, and a deliberate but prompt investment in training, equipping, and organizing the forces needed to protect our nation in the 21st century. The "2+" strategy of structuring the force to address two major theater war (MTW) scenarios at once is driving our military into severe degradation. The Revolution in Military Affairs (RMA) is not a substitute for strategy and it is bankrupting our military by diverting what disposable funds we have toward an overly technical "system of systems" that is neither financially nor militarily sound. At the same time, RMA is creating an enormous interoperability gap—a strategic deficit—between our forces and those of allied nations, and between our commanders and the 98 percent of the relevant information they need that is in the private sector and not accessible by our intelligence, surveillance, and reconnaissance (ISR) systems.

This review, after evaluating the real-world threat, outlines a change in our national security strategy from 2+ to 1+iii—we need *four* forces after next, not one—and an increase in national security spending on the order of $40 billion a year for traditional military capabilities and $10 billion a year for nonmilitary capabilities in direct support of our long-term national security strategy. Regardless of funding, however, we need to restructure the force.

Arriving at the Bottom Line Figure.

Senator Sam Nunn, then Chairman of the Senate Armed Forces Committee, said in the 1990s, with perfect clarity:

> I am constantly being asked for a bottom-line defense number. I don't know of any logical way to arrive at such a figure without analyzing the threat; without determining what changes in our strategy should be made in light of the changes in the threat; and then determining what force structure and weapons programs we need to carry out this revised strategy.

This review follows Senator Nunn's cogent tasking by first discussing the threat, then recommending a strategy appropriate to the threat, and finally proposing specific force structure modifications as are necessary to execute the new national security strategy, a strategy I call the "1+iii" (One Plus Triple I) Strategy. This new strategy will reinforce our conventional military; substantially enhance our expeditionary, constabulary, and special operations forces; create a bold new program to achieve force protection through global intelligence coverage that inspires economic and cultural investments; and assure home front security through a much expanded and better integrated combination of electronic security and economic counterintelligence that extends the concept of national security down to the state and local level through revolutionary new uses of our National Guard and Reserve forces.

Analyzing the Threat.

The "threat" to the United States in the 21st century must be evaluated in the larger context of a world where conflict is the norm, where major ethnic fault lines cut across all major continents, where transnational criminals and local warlords are amassing fortunes through trade in women, diamonds, food, and medicine; and where water—our most precious resource—is approaching a "tipping point" of nonrenewability.

Let us begin with conflict. Each day, today, we have on-going 26 severe low-intensity conflicts that killed over 300,000 people in 1999 alone, and cumulatively, have killed roughly 8 million over time. There are 78 less severe low-intensity conflicts, and over 178 violent political conflicts internal to specific nation-states. India, Nigeria, Indonesia, Pakistan, Colombia, China, Russia, Uganda, Ethiopia, and Sudan, all populous countries, are engaged in between 6 and 32 conflicts each!

Conflict trends are troubling. Severe low-intensity conflicts (defined as conflicts with over 1000 casualties per year), have leveled off. However, lesser low-intensity conflicts are increasing steadily in number each year, while violent political conflict, often ethnically-based, has leaped toward geometric increases year by year. Figure 1 shows the actual number of conflicts per year from 1995 to 2000.

In addition, relying on the aggregate data collected and analyzed by centers of excellence such as the Stockholm International Peace Research Institute (SIPRI), we see that our world, today, endures 29 complex emergencies as declared by the United Nations; millions of refugees and internally-displaced persons across 67 countries; food scarcity and related disease in 27 countries; modern plagues, from AIDS to the West Nile disease, creeping across 59 countries and rising; and child soldiers murdering one another in 42 countries. Peacekeeping forces are in 38 countries; landmines desecrate 62 countries; torture is common in 92 countries; corruption is common in 78 countries; and censorship is very high in 63 countries.

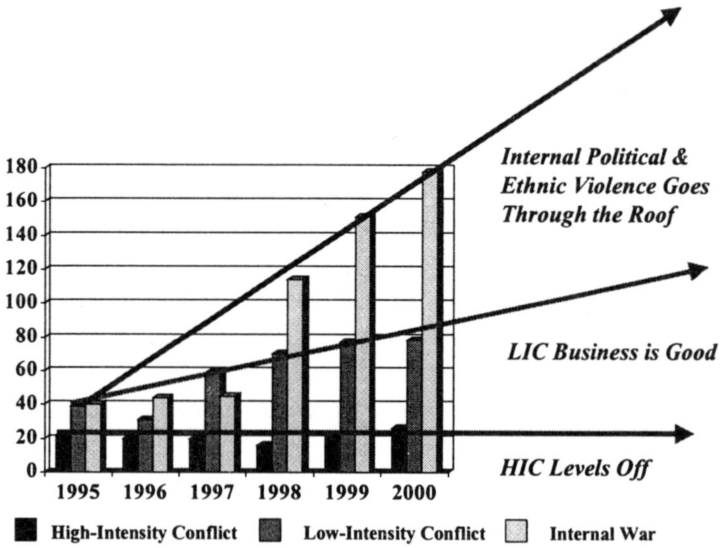

Figure 1. Conflict Trends from 1995-2000.

Those are simply the conflicts and the obstacles to effective government management of scarce resources on behalf of their people. Let us turn to the special cases of ethnicity and water. Ethnicity, despite the popular case made for a "clash of civilizations," is really most relevant when it is combined with desperate shortfalls in the basics of life, such as water. Figure 2 combines a map of the current state of water for the world with genocidal fault lines corresponding to major ethnic divisions.

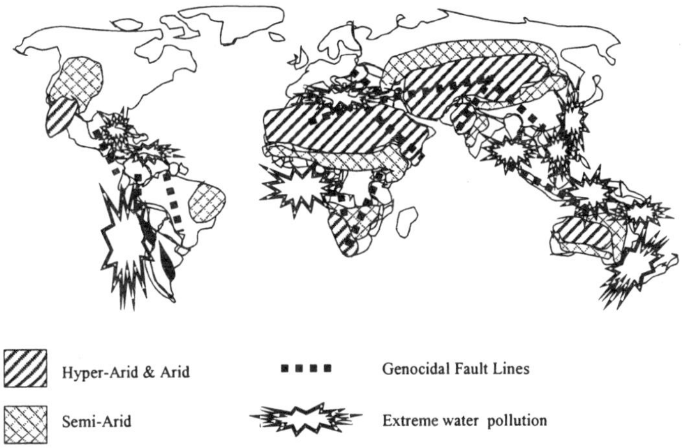

Figure 2. Intersection of Water Scarcity and Genocidal Fault Lines.

The coincidence of water scarcity and ethnic fault lines in the Slavic-Islamic and Slavic-Chinese border regions is of special concern. Closer to home, we must be conscious of both the increasing hyper-aridity and declining aquifers of the American mid-west, and the substantial pollution characterizing all of the Gulf of Mexico and the Caribbean.

The greatest threat to both national security and national prosperity in the 21st century stems from a combination of water scarcity, failed states, ethnic fault lines, and opportunistic thugs thriving under conditions of chaos. We are close to a "tipping point," and it is we who are creating the ultimate crisis that results from a combination of global water pollution and the degradation of flood plains (no longer receiving nutrients because of dams blocking the silt) and the effects of irrigation (raising the salinity of soil to a point where it cannot produce food) and vanishing aquifers (being mined into extinction); with genocidal fault lines and the attendant instability that gives rise to rogue warriors.

Our national intelligence communities, while focusing primarily on strategic nuclear and conventional threats and those aspects of the threat that are secret, are fully aware of these dangers, but unable to make a compelling public policy case for action. The Central Intelligence Agency (CIA) did an excellent job of forecasting the spread of Anti-Immunal Deficiency Syndrome (AIDS) in the 1970s, but the policy community was not willing to make this an international issue nor to allocate resources for preventive measures. More recently, Dr. John Gannon, Associate Director of Central Intelligence for Analysis and Production (ADCI/AandP), has rather carefully pointed out that the major threats facing us in 2015 are related to mass migrations, disease, and other nontraditional factors. Despite a major news story on the gap between intelligence warning of AIDS in the 1970s and policy action on AIDS for a quarter century thereafter, Dr. Gannon's accurate and timely warning about emerging nonmilitary threats is being ignored.

At the same time, selected experts and the occasional rare reporter have begun to focus on "modern plagues" as well as water shortages, but they do so only within their professional circles and fail to get a hearing at the policy level. Even those books that receive presidential and broadcast television endorsements, such as Laurie Garrett's *Betrayal of Trust: The Collapse of Global Public Health* (Hyperion, 2000), fail to affect the national and other state budgets for the simple reason that the voters—the citizens—will not buy a 754-page book, much less read it, and still less act upon its well-documented and urgent message. The heart of Garrett's message merits our attention.

In this context we must acknowledge the importance of the new definition adopted by the United Nations in Security Council Resolution 751 of April 24, 1992, where the "magnitude of human suffering" in Somalia was recognized as constituting a threat to peace and security. We do this for two reasons: because such suffering creates waves of migration that carry disease, and because our "home defenses" against epidemics have been allowed to atrophy to the point that we are at serious risk in the developed world and at the provincial, state, and local levels.

The threat in the 21st century is more complex than ever before and cannot be defined in strictly military terms. Figure 3 provides a means of understanding this complexity while structuring the threat in a manner that leads logically to both strategic and force structure equivalencies.

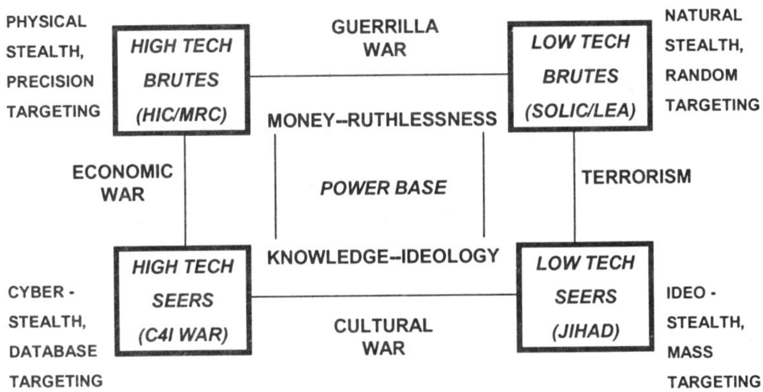

HIC	High Intensity Conflict	SOLIC	Special Operations/Low Intensity Conflict
MRC	Major Regional Conflict	LEA	Law Enforcement Agencies
C4I	Command, Control, Communications, Computing and Intelligence	Jihad	Religious conflict, also includes environmental, non-state "war"

Figure 3. Four Threat Classes Requiring Strategic Consideration.

In fact we face four general kinds of threats: the traditional nuclear and conventional forces sponsored by a state; those that are violent but not necessarily associated with a state, including both transnational criminals and terrorists or warlords able to acquire weapons of mass destruction; those that are nonviolent and often stateless, including environmental conditions imposing a high "magnitude of human suffering" as well as the refugees—often gravely ill—from those conditions, the child soldiers bound into armed slavery, and the women and children traded for money and often laden with disease; and finally those threats to home defense, be they state-sponsored or not, that surround our critical infrastructures, including our public health infrastructure, and the core of our economic well-being. At times, it is ourselves that we have to blame for the scope and imminence of our vulnerability, as is the case with public health.

Seen another way, these four threat classes confront us with four distinct "ways of war": Systemic War, Dirty War, Peacewar, and Cyberwar. Further complicating our planning and programming, conflict between differing forces takes differing forms, and we must evaluate how they fight and how we might fight in the context of a world that does not favor heavy armor formations—a world in which only 50 percent of the ports are usable, where there is almost no cross-country mobility, bridge loading is limited to 30 tons and less in most Third World countries, and the aviation climate is hot and humid.

Changing the Strategy.

Fundamental strategic thinking should include an appreciation for the fact that a national security strategy must be *holistic*—managing all sources of national power including diplomacy, economic assistance, cultural outreach, and information operations, not just the military—simultaneously. "War proper" is not just about military force, but rather about imposing one's will and assuring one's security in a complex world. Within this larger context, power without purpose is wasted, time is priceless, technology is not a substitute for strategy or thinking, asymmetric threats must receive co-equal attention with symmetric threats, and strategic culture *matters*.

Determining our national security strategy for the 21st century therefore must be guided by two related principles: co-equal standing for asymmetric versus symmetric threats; and co-equal structure and funding, or at least some semblance of a rational balance, between *military* forces designed for the traditional symmetric threat, and largely *unconventional or nonmilitary* forces designed to deal with the asymmetric threat.

On this basis, "forward engagement" and "shaping" of the theater environment make a great deal of sense, but with two enormous caveats: there must be a force structure as well as funding for nonmilitary investments, and we are probably better off talking about "nurturing" peaceful environments instead of the more imperial "shaping." At a minimum, a strategy that is seriously committed to force protection through economic, cultural, and information peacekeeping must recognize the vital role played by the nongovernmental organizations (NGO); the critical importance of being able to communicate and cooperate with indigenous organizations that are not part of a military force; and the overwhelming influence on any situation of environmental conditions including the availability of clean drinking water, sufficient food for the children, and such medical provisions as might be needed to at least keep disease from spreading through epidemics.

Our new national security strategy must actually have five elements that are in complete harmony with one another: our *global intelligence strategy*, for ensuring that we can maintain global coverage and global warning; our *interoperability strategy*, for ensuring that what we build and buy is interoperable with both military and civilian coalition partners in a wide variety of "come as you are" circumstances; our *force structure strategy* for ensuring that we build to both the most likely as well as the worst case threats while balancing the relative roles of our military, the rest of the Federal government, the reserve force, the private sector, and external allies or coalition partners; our *preventive diplomacy strategy* for directly addressing conditions around the globe that spawn conflict and crises; and finally, our *home front strategy* for fully developing and integrating the defensive capabilities of our state and local governments and the private sector.

A truly "transformative" defense strategy would recognize that in this complex world with four threat classes we must adopt a "total mobilization" approach to national security, and ensure that every element of government at the federal, state, and local levels is empowered and integrated into an effective "total force" while we also ensure that the private sector is doing its part, particularly in relation to documenting supply-chain vulnerability for high-technology forces and in applying new "due diligence" electronic security measures to raise the over-all security of our national financial, communications, power, and transportation infrastructures. Figure 4 illustrates the kinds of trade-offs that must be made if we are to have a "transformative" force structure strategy.

Without spending too much time on these trade-offs, let us just note that there are three kinds of trade-offs shown in the figure: between the military and the rest of government; between the active force and the reserve force; and between the government as a whole and the private sector. We will leave the issue of U.S. versus allied or U.S. versus NGO coalition levels of effort for another day.

"Real WAR" (HIC/MRC) forces must protect the core military and rely almost completely on active duty personnel "ready to go" without waiting for reservists; and will draw on private sector capabilities to the minimal extent possible.

"WAR"	SOLIC/LEA
80% Military – 20% Other 75% Active – 25% Reserve 75% USG – 25% Private	50% Military – 50% Other 50% Active – 50% Reserve 50% USG – 50% Private
HOME FRONT	**PEACE**
20% Military – 80% Other 50% Active – 50% Reserve 25% USG – 75% Private	40% Military – 60% Other 25% Active – 75% Reserve 50% USG – 50% Private

Figure 4. Transformative Force Structure Trade-Offs.

SOLIC/LEA forces, by contrast, will see the U.S. Government (USG) fielding an even mix of military and diplomatic or justice or economic capabilities, while also drawing equally on active and reserve forces, and dividing the responsibility for dealing with terrorism and transnational crime equally between U.S. Government endeavors and private sector security and intelligence activities.

In peace, the military continues to provide a global logistics and communications infrastructure, but civilian elements of the U.S. Government are in the majority role. Reservists skilled at foreign languages and with occupations vital to civil affairs and the restructuring of failed states come to the fore, while the overall effort is balanced between USG-funded and manned activities, and "overt action" by private sector elements including NGOs.

Finally, for IO/ECON, there remains a 20 percent commitment of military forces—largely in the National Security Agency (NSA) and related service information warfare centers—while the Justice, Treasury, and other departments come to the fore; there is an even split between active duty forces carrying out Information Operations duties, and elements of the National Guard carefully positioned across all critical infrastructure nodes, with the funding—and the ultimate responsibility for day-to-day security—resting primarily with the private sector.

On the basis of this kind of approach, one can readily validate a need for four regional Commanders-in-Chief (CINCs) (Pacific, Southern, European, and Central) while conceptualizing four "threat-type" Commanders-in-Chief (WAR, SOLIC, PEACE, and HOME). It would be these eight CINCs that should comprise the working level of the new Joint Requirements Board under the direction of the Joint Chiefs of Staff.

Such a force structure strategy would at a minimum restructure the relationships between the Departments of Defense, Justice, and State; would establish minimal mandatory defense structure needs within the Departments of Commerce, Treasury, and Transportation as well as the Federal Reserve; and would create selective new relationships—including secure interoperable communications networks—with state and local agencies, the Immigration and Naturalization Service, the Federal Emergency Management Agency (FEMA), and such other civilian elements of government as must be better integrated into our "total force" strategy. The president's immediate staffs—the National Security Council and National Economic Council and other odds and ends—should also be restructured to conform to the need for matrixed management of integrated operations against each of the four threat classes. Such a force strategy would also establish, in clear terms suitable for a news media report as well as legislation, the minimal mandatory responsibilities of the private sector in support of our national security strategy, with a special emphasis on very high standards for electronic security.

An integrated national security strategy, then, must carefully develop, in tandem and with appropriate fiscal resources as well as force structure being assured for each element of this *holistic* strategy, each of the following: a global intelligence strategy; an interoperability strategy; a force structure strategy; a preventive diplomacy strategy (including economic assistance and cultural programs); and a home front strategy. Those who would persist in limiting our national security strategy emphasis to conventional military forces are demeaning Clausewitz and undermining the security of the nation.

Determining the Force Structure.

A national security strategy that addresses all of these factors, with all of the legal, financial, and political implications that are associated with different kinds of "engagement," must inevitably find that we need *four* forces after next, not one.

While it is certainly possible to have one "core force" that includes the world-wide mobility, logistics, and communications capabilities that we are justly proud of, in fact our strategy must find that we need:

• a nuclear and conventional force that is smaller but very well equipped, fully modernized, and never committed to OOTW—the WAR force;

• expanded and enhanced expeditionary, constabulary, and special operations forces able to put increasing force packages anywhere in the world within 24 to 48 to 72 hours—the SOLIC force including direct support to LEA;

• a PEACE force, possibly combining substantial elements of the Civil Affairs, Army Engineers, and the Agency for International Development with new liaison elements specially trained to interact with civilian rescue units, as well as a new humanitarian assistance fleet within the U.S. Navy and also new Air Force lift capabilities relevant to peacekeeping; and finally,

• a fully developed HOME Defense force that gives state and local authorities, not just federal authorities, everything they need to legally carry out their duties in preventing economic espionage and electronic attack against any of our critical infrastructures, while integrating the U.S. Coast Guard and appropriate national missile defense and other "continental" defense capabilities.

We cannot rely any longer on just the military, or on a "one size fits all" military where our people and equipment are assigned to all kinds of missions for which they have not been trained, equipped, and organized.

The "Core Force," as opposed to General Colin Powell's "Base Force" approach, draws a distinction between core functionalities and capabilities that are needed for a global presence—communications, logistics, mobility, manpower management—and very distinct and carefully focused force structures and organizational arrangements that are self-sustaining and are very deliberately trained, equipped, and organized for optimal effectiveness in one of the eight "core competency" areas shown in Figure 5 below.

Figure 5. "Core Force" Visualization.

Eight Functionalities, Four "Type" CINCs.

Each of these eight functionalities should be actualized in corresponding force structure initiatives.

Strategists. Our national Net Assessment capability, and our national as well as our military strategic formulation processes, have broken down. They have become bureaucratic exercises of little value to long-term force structure planning. They are weak in part because no one has been willing to challenge the many false assumptions and premises that guide our current force structure decisionmaking process. We need, at a minimum, a dedicated National Security Strategic Center that has an even mix of representatives from each of the major slices of national power, as well as an even mix between long-term strategic thinkers specializing in each of the four threat classes, and "top 5 percent" personnel from the military, other elements of the civilian government, state and local law enforcement and public health, and the private sector, with special regard for selected nongovernmental and nonprofit sectors. This element should report directly to a new Presidential Council but be managed on a day-to-day basis by the Joint Chiefs of Staff.

Domestic Threat. FEMA has improved incrementally in recent years, but needs a great deal more authority and financial support. We need to redirect a substantial portion of the National Guard toward national, state, and local emergency response duties, and to give them the training, equipment, and organization that they require to become extremely effective at dealing with fires, riots, and epidemics, in direct support of the constituted legal authorities being assisted. From communications to medical to civil engineering to public relations to food services, there are valid requirements that demand a "total make-over" for those elements of the National Guard fortunate enough to be selected for this very urgent and honorable aspect of national defense. This force, to include new investments in active duty personnel as a cadre and law enforcement specialists as well, should be under a "type" CINC for Home Front Defense who would also be responsible for Electronic Security and Citizen Education as discussed below, as well as for the national missile defense system as it develops over time.

Force on Force. This is the traditional military, responsible for creating the maximum amount of violence in the smallest possible space—responsible for being able to execute "scorched earth" missions that obliterate entire cities if necessary, that can control significant areas of terrain in order to find and kill exactly the right key personnel threatening the United States with anything from transcontinental missiles to bio-chemical car bombs. This force must receive all that the RMA can offer it, while also being protected from OOTW missions and other distractions. This force, under the leadership of a specific CINC responsible for the "total war" mission, should take over the bulk of the existing defense funding, and focus exclusively on maintaining its readiness while modernizing aggressively. This force must have air-ground task forces dedicated to specific regions of the world and at least one complete Corps specifically trained for each of the four major terrain types over which major wars might be fought: desert, jungle, mountain, and urban. Selected elements of all of the other forces (e.g. small wars, constabulary) would pre-plan and train for specific contingency missions in support of a Major Regional Conflict (MRC) campaign, and in the event of an MRC, would be chopped as required to the operational control of the regional CINC who would also receive operational control of force on force elements.

Small Wars. We keep forgetting our history. Both the British and the Americans have learned the same lesson more than once: forces designed for traditional conflicts do not do well in small wars until they have undergone such considerable adaptation as to render them unprepared and ineffective when required to return to traditional warfare. Small wars require a much higher standard of foreign area knowledge and language competency, to name just one significant difference, and are best fought by units trained, equipped, and organized specifically for small wars. The Special Operations Command (SOCOM) is ideally suited by both its strategic culture and its tactical excellence, to serve as the parent of a force of three division-wing teams optimized for expeditionary operations. This force would be especially skilled at joining international coalitions engaged in peace *enforcement* operations, and in executing violent complex forced entry missions.

Constabulary. The force that fights the small wars is *not* the force best able to maintain the peace, restore the functions of the failed state, and generally move as quickly as possible toward an exit that has been planned by the original engagement strategy. Constabulary forces require a combination of enormous numbers of civil affairs personnel, very high percentages of military police, engineering, medical, and food service personnel, and considerable communications, intelligence, and liaison personnel. This force must draw on and implement major civilian programs related to water purification and desalination, food purity and distribution, and epidemic conditions. This force must excel at working with and sustaining long-term relations with NGOs—a major challenge where our traditional intelligence and operations leaders have failed completely. It has to fully integrate indigenous personnel into every aspect of its reconstruction of society and the eventual turn-over of authority to indigenous leaders. It has to provide considerable training in many skill areas, and at the same time needs to plan for a deliberate abandonment of most of its equipment, including communications equipment, as part of "the deal." This force would then return to the Continental United States (CONUS) to reconstitute itself. It should be under CINC SOLIC.

Ground Truth. The average Embassy officer is not trained, equipped, nor suited by nature to go in harm's way on a daily basis. Most of what we need to know in the Third World is not published at all, much less in digital form or in

English. Our increasingly complex world requires that we have a force for establishing "ground truth" through direct personal observation, in every clime and place. Such a force, created in the defense attaché mode but with much greater freedom of movement and much deeper mobility and communications support, would integrate overtly as assigned liaison officers; "circuit riders" assigned to entire countries or regions and told to stay out of the Embassy and off of the cocktail circuit; and very selective networks of clandestine and covert observers using third party passports or surreptitious entry to obtain their direct "ground truth" observations including sensitive measurements and signatures intelligence (MASINT). This force should be under CINC SOLIC, but in keeping with my recommendations for overall intelligence reform there should be a major Clandestine Services Agency (CSA) Station co-located with CINC SOLIC to ensure optimal coordination between these "early warning" observations by warriors focused on rural areas, and the more traditional civilian clandestine espionage activities focused on urban political and economic and military targets.

Electronic Security. We have a very long way to go before it is truly safe to live and work in cyberspace. Our financial, communications, power, and transportation infrastructures remain totally exposed and vulnerable for the simple reason that we will not be secure until there is a wholesale conversion of all existing electronic systems to a high level of security that must be embedded from the factories of the components on out. This will require three major national initiatives that are not yet being properly discussed in Washington: 1) the definition of minimal mandatory standards for hardware, software, and personnel security in relation to electronic systems and their contents; 2) the imposition of these standards via "due diligence" legislation that requires all enterprises to be compliant within 5 years, with some systems to be secure within the year; and 3) the complete release of NSA-level encryption to the private sector so that the Internet can be as secure as Presidential communications. This level of security on the Internet is in fact a precursor to enabling the intelligence community as well as corporations to have access to *all* relevant information while still being able to process secrets. The minuscule effort being made today must be supplanted by a trained force responsive to CINC HOME, and electronic security brigades specializing respectively in financial, communications, power, and transportation systems, that are in turn integrated within a nationally distributed "virtual network" of private sector employees, National Guard specialists, and Home Front Force active duty cadre in new consolidated electronic operations centers focused on each of the major systems areas.

Citizen Education. "A Nation's best defense is an educated citizenry." As Senator Boren and David Gergen have noted so clearly, we are in fairly desperate circumstances in relation to both policymaker and voter knowledge about the hard realities of the world we live in. A major investment must be made in the "internationalization of education," but even more so, we must find ways to better integrate our increasingly diversified population so as to create a minimal level of social cohesion over time. It is my view that we must restore the draft and require every U.S. citizen to serve for 4 years, in any combination of years (e.g. 2 + 2 or 1 x 4) between their 18th and 38th birthdays, with at least basic training and the first year being required before entering college. I must go further and in recognition of both the middle-aged immigration increases as well as the longevity increases, and say that we must have an additional draft requiring 2 years of service (at once or in stints of 3-6 months) from all those who are citizens in their 38th to 58th years and have not served previously. At the same time, we must substantially increase private sector sabbaticals by our field grade officers and selected senior non-commissioned officers. We must, in effect, give true meaning to the concept of "total force" by ensuring that every single citizen has a common foundation of service to the nation, and that we fully integrate every citizen—to the extent of their capabilities—into our national defense. This educational process does not require that every citizen bear arms—our *new* national defense force structure will offer many opportunities for those who do not wish to wear a uniform or learn how to kill.

In summary, all but one of these eight functions would be integrated under one of four "type" CINCs that would in turn support the regional CINCs much as the services do today but with a vastly improved *focus of effort* that assures both air-ground-sea interoperability as well as joint training and doctrine suited to the specific "type" of warfare to be fought. Over time we should convert each of the four Services into one of the type CINCs, or disband them as we downsize administrative capabilities and improve our tooth-to-tail competency under this new force structure approach.

Reordering the Government.

As Admiral William J. Crowe, Jr., has noted, we need to do a much better job of organizing the rest of the government so that it is capable of "forward engagement" using *all* of the sources of national power. This should require, at a minimum, the establishment of Ambassadorial-level appointments to each regional CINC from Commerce, Treasury, the Peace Corps, and the Agency for International Development, and up-grading of the existing Political Advisor positions from State to Assistant Secretary-equivalents.

A new CINC PEACE, as a "type" CINC, should be established with a small staff in the National Capital Area, close to the Department of State, and able to draw on military command and staff personnel as well as military dollars to ease the transition toward the day when we have proper funding and structure for the nonmilitary elements of national power. Someone like General Colin Powell or Admiral William Crowe, Jr., would be ideal candidates to serve as CINC PEACE, with *international* education, water, food, and public health as the mandated areas of interest, and the right to cut across bureaucratic boundaries, on behalf of the President, when it makes sense to do so. Eventually, once we have our national security house in order, we will find that a similar restructuring of government is necessary with respect to health, education, interior, and other domestic elements of government responsible for the internal "commonwealth."

CINC WAR	CINC SOLIC	CINC PEACE	CINC HOME
Force on Force	Small Wars	State/USIA	Domestic Threat
	Constabulary	Peace Corps	Electronic Security
	Ground Truth	Economic Aid	Citizen Education

Figure 6. Reconstitution of Force Structure.

Some of the features of each of the ten force structure components of a new national security "total force" are shown below.

CINCWAR

- **Force on Force**
    - Existing strategic nuclear forces, drawn down as appropriate, but modernized
    - Four Army-Air Force Corps of 3 division-wing teams each (12 and 12)
    - Strategic mobility (black and gray) to move one corps in 4 weeks

CINCSOC

- **Small Wars**
  - Complete absorption of the U.S. Marine Corps, without dilution of its Congressionally-mandated character or culture including uniforms
  - Implementation of the 450-ship Navy (rapid response, littoral squadrons)
  - Creation of two active/ four *Reserve* foreign area combat support brigades
- **Constabulary**
  - Five active and five *Reserve* constabulary task forces
  - Implementation of the 450-ship Navy (humanitarian assistance slice)
- **Ground Truth**
  - Ten *Reserve* foreign area specialist companies
  - Ten *Reserve* ground sensor /relay communications platoons (covert capable)

CINCPEACE

- **State/USIA**
  - 1000 additional foreign service/foreign information officers
  - 100 new consular/open source information posts
- **Peace Corps**

- 10,000 new Peace Corps volunteers per year
- Implementation of Peace Corps information assistance program

- **Economic Aid**
  - 10 new water, food, and medicine projects each year (2 within USA)
  - Digital Marshall Plan for the Third World

CINCHOME

- **Domestic Threat**
  - 50 National Guard Brigades, each with fire, riot, engineer, and medical battalions that train with state and local counterparts and also do international humanitarian assistance and disaster relief
  - National missile defense for New York and Washington as soon as possible
  - Absorption of the U.S. Coast Guard, without dilution of its character
- **Electronic Security**
  - 50 National Guard Electronic Security Centers
  - 50 National Guard Electronic Security Battalions, with specialist companies
- **Citizen Education**
  - University of the Republic

- Universal Draft/National Defense Fellowship Program

This approach to managing how we train, equip, and organize our varying force structures to deal with four distinct threat challenges wipes out, in one grand "Goldwater-Nichols"-style revolution, all of the negatives of the existing Service "stovepipe" acquisition systems and the cultures that go with them. This should be the focus of the National Security Act of 2001.

Conclusion.

Both the RMA and the so-called defense transformation movements have failed. The 2+ MRC strategy has failed. Our security environment demands a Home Front Force; a Peace Force; a Ground Truth, Small Wars and Constabulary Force; and a dedicated strategic nuclear and conventional War Force that is not frittered away on OOTW.

We require a National Security Act of 2001. This strategy, and the attendant force structure, are achievable within 6 years from where we are today, but will *not* be achievable as readily if we delay because the U.S. Navy is decommissioning ships as we speak—we must put a stop to their dismantling of our submarine, destroyer, and frigate capabilities because it is the U.S. Navy, as CINC PEACE, that will have the greatest burden to bear in support of CINC SOLIC (U.S. Marine Corps) and CINC WAR (U.S. Army). We must give CINC HOME (the U.S. Air Force) the financial resources—and culturally-powerful incentives—with which to rapidly reconfigure itself into an effective Home Front Defense that fully integrates and respects the needs and concerns of our state and local and private sector partners in our "total war" environment.

If we adopt a 1+iii strategy and implement the recommendations of this review, America will begin the 21st century with a national security architecture well-suited to our needs and agile—able to fight and win in any clime or place.

CHAPTER 10

COMPREHENSIVE SECURITY AND A CORE MILITARY CAPABILITY

Walter Neal Anderson

> Despite the proliferation of highly sophisticated and remote means of attack, the essence of war will remain the same . . . it will not be like a video game. What will change will be the kinds of actors and weapons available to them. While some societies attempt to limit violence and damage, others will seek to maximize them, particularly against those with a lower tolerance for casualties.
>
> Hart-Rudman Commission
> *New World Coming*, September 1999

Introduction.

At the end of the Cold War, the U.S. military moved from a force focused on fighting the Soviet Union to a force capable of and sized to perform a broad range of missions against a backdrop of an uncertain future. America's armed forces are now ostensibly structured to fight two Major Theater Wars (MTWs) nearly simultaneously. It is not surprising, however, that although reaffirmed by the 1994 Bottom-Up Review and 1997 Quadrennial Defense Review (QDR), there has been much discussion recently about the utility of the so-called "two MTW strategy." Many, such as the Hart-Rudman Commission, see it as antiquated and no longer relevant.

Despite a dearth of substantive intercourse on our national defense strategy during the recent presidential campaign, events have converged to make a thoughtful debate over America's security particularly timely. While there is no single threat on which we can or should focus, the end of the Cold War, the beginning of a new millennium, a new administration, and the mandated QDR process give us ample reason to take a fresh look at the issues affecting the security of our nation. At the end of this debate, the strategic path we take will determine the well-being of all Americans for generations to come. Thus, it is imperative that the decisions we take be as well considered as possible. However, we will do it wrong if we start the discussion with the "two MTW strategy," for it is a misnomer of the highest order.

America's national security strategy is currently one of *Engagement*, and the military has put it into practice via its national military strategy of *Shaping* the international security environment and *Responding* to threats and crises, while *Preparing* for an uncertain future. While the label of "Engagement" likely will be revised by the new administration, America's security strategy is unlikely to change drastically. The military will continue to perform *Shaping, Responding,* and *Preparing* missions, even if the names of and relative emphasis on these three elements change. Indeed, that, to date, the debate on our defense strategy really has been over the balance between these three strategic fundamentals is testimony to the continuing role they likely will play in our national security. As we reexamine this paradigm, we must consider our enduring national interests and responsibilities, as well as the broad strategic and operational requirements and potential challenges on the horizon.

Enduring Global Interests and Responsibilities.

Our strategy debate must start with national interests. The Clinton administration classified our interests as vital, important, humanitarian, and other. While some would offer new categories, most agree that our interests should be clearly articulated and prioritized. Although our national strategy and the importance of various issues may have changed over the years, America has always embraced three fundamental interests:

- Preserving the sovereignty of our nation, with values, institutions, and territory intact;
- Protecting our citizens at home and abroad; and,
- Providing for the common welfare and economic prosperity of our nation and citizens.

These enduring interests rightfully drive our national security objectives and attendant military capabilities, and have led to a broad consensus that the United States must remain actively engaged in the world. Isolationism or retrenchment is not an option. Globalization punctuates the fact that our interests are undeniably tied to actors and forces beyond our borders.

Because we have entered the new century with unprecedented national strength and extraordinary economic health, the military's role regarding the third interest—common welfare and economic prosperity—is worthy of particular note. While the military's influence on physical security is obvious, many Americans do not appreciate the connection between our national prosperity and our military excellence. Although physical security may not be sufficient for prosperity, it is absolutely necessary.

Surveying the past decade, we now see that the real peace dividend following our Cold War "victory" was far less the result of reduced military expenditures than accelerated prosperity resulting from the diminished threat of global war, which allowed most countries to focus on economic development. The global contest between capitalism and communism was swept away as democratic, free market countries found themselves on the right side of history. Security was the oxygen that allowed the global economy to breathe—peace, stability, and confidence were essential for the sustained prosperity we have enjoyed over the past several years. This will be even more true in the future, as the roots of global economic interdependence broaden and deepen.

America also has enduring global responsibilities. Formal agreements and informal friendships around the world inextricably tie U.S. security and well-being to critical allies' and partners.' While these relationships do serve others' interests, our motive is not beneficence. We maintain and nurture these relationships because it serves America's interests to do so. The United States would be hard-pressed to "go it alone," and has no desire to pursue unilateral solutions for global issues. Thus, as background for our debate, we must keep in mind the potential implications of any strategy shift that would signal to either allies or potential adversaries that America is retrenching. There has been, for decades, some tension between the United States and its allies over mutually satisfactory levels of security "burden sharing." Often this tension has had a negative effect on overall relations. For its part, the United States is seen by some as "wanting it both ways"; as wanting leadership of an alliance or international body without paying our full "dues" in terms of participation, sacrifice, and commitment. Clearly, each case must be judged on its own merit, but America must work closely with its international partners to strike an appropriate balance. Leadership comes with a cost. If we are willing to commit less and relinquish leadership in some cases, we must be prepared to accept the consequences of diminished influence.

Taken together, advancing America's enduring global interests, fulfilling its worldwide responsibilities, and sustaining its influence have a price tag. Any hope that the end of the Cold War meant the "end of history" remains unfulfilled. Instead, America's armed forces, at one-third their Cold War size, are more heavily committed globally than ever. As the new administration begins to keep its promise to review America's current global engagement posture, it must acknowledge the evolving strategic landscape and the increasingly complex and diverse challenges to America's security and well-being.

Strategic Challenges and Shifting International Sands.

Strategic challenges to U.S. interests have evolved dramatically in the last 10 years. Despite positive economic and political trends since the end of the Cold War, not all states have benefited equally from these changes. The world order is in transition. China, Russia, India, and Indonesia, comprising more than a third of humankind, are in the midst of major, uncertain political, economic, and social change. Released from the grip of a bipolar structure, state and nonstate actors can more easily challenge the global distribution of power, the concept of sovereignty, and the character of warfare. Traditional friends and allies are rethinking the global security environment and their place in it. Even Europe is changing as it seeks ways to further integrate, while protecting individual national identities. Real or invented notions of history, ethnicity, and religion are seemingly more evident. The Balkans and other regions remain frozen in a state of "no war, no peace." We have yet to even name this era, and its ultimate direction and outcome are anything but certain.

As we negotiate shifting international sands, there are three enduring truths. First, the United States has, and would like to maintain, *comprehensive preeminence*—politically, economically, and militarily—albeit less arrogantly. Second, power remains relative, dynamic, and multidimensional. And third, armed conflict—rooted in the contest for power and control over people, territory, and resources—remains a permanent feature of the global landscape. These truths have direct implications for our security debate.

Even though general war with a global competitor or hostile alliance appears unlikely for now, there are many broad categories of threats that confront us. Challenges likely requiring a military response include:

- Humanitarian disasters, man-made or natural;

- Transnational threats, including terrorist groups, illegal drug trade, international organized crime, and piracy;

- Failed states and intra-border conflicts, where political, ethnic, religious, or economically driven violence threatens broader instability;

- Attacks on the United States homeland or on Americans abroad; and,

- War, including cross-border aggression, where states threaten the sovereignty of others in regions central to U.S. interests.

The character of these threats is also changing. Potential adversaries have taken to heart the lessons of 20th century conflict, including Operation DESERT STORM. They are selectively modernizing their militaries and developing both traditional and asymmetric capabilities to avoid or counter our strengths. These changes have significant implications for the environment in which the United States and its partners might operate. State and nonstate adversaries likely will:

- Employ chemical, biological, nuclear, or radiological weapons;

- Exploit real or perceived U.S. political, economic, cyber, or psychological vulnerabilities;

- Operate in complex and urban terrain, hiding behind innocent civilians; and,

- Delay, disrupt, or deny U.S. military response and access through conventional or unconventional tactics or capabilities.

An understanding of these emerging challenges must inform our strategic debate by giving us insight into our defense strategy and the military capabilities we will likely require to implement it.

Enter the Debate.

Our current national security strategy serves as a useful point of departure for our debate. As mentioned previously, it is a strategy of "Engagement" and states we must *Shape* the security environment, *Respond* to threats and crises, and *Prepare* for an uncertain future.

Our current *National Security Strategy (NSS)* also stipulates that:

> [America's armed forces] must have the capability to deter and, if deterrence fails, defeat large-scale, cross-border aggression in two distant theaters in overlapping time frames. Maintaining a two major theater war capability reassures our allies and friends . . . [and] deters opportunism . . .[1]

From this portion of our current security strategy the two MTW force-sizing standard—first established in 1991, and reaffirmed many times since—is articulated.

The *NSS* continues by noting that:

> Fighting and winning major theater wars entails three challenging requirements. First, we must maintain the ability to rapidly defeat initial enemy advances short of the enemy's objectives in two theaters, in close succession. . . . Second, the United States must be prepared to fight and win under conditions where an adversary may use asymmetric means against us. . . . Third, our military must also be able to transition to fighting major theater wars from a posture of global engagement—from substantial levels of peacetime

engagement overseas as well as multiple concurrent smaller-scale contingency operations.[2]

Much of our current risk originates from the "three challenging requirements." As we discuss any future strategy, we must address these issues candidly.

Concerns over the Current Strategy—Terminology and More.

Our current definition of a Major Theater War as "large-scale, cross-border aggression" is too narrow to address current and future strategic realities. First, the scenarios for MTWs are typically linked to the Korean Peninsula and Persian Gulf. Clearly the possibility exists that the United States will find itself involved in a war beyond these two contingencies. Second, it is possible to fight a war without "cross-border aggression" by an adversary. Operation ALLIED FORCE illustrates both of these points. This operation was, for U.S. Air and Naval Forces, as well as critical enabling capabilities like C4ISR and strategic lift, "MTW-like" in the resources required. Air Force officials repeatedly made this point. Had a forced entry into Kosovo been necessary, the ground forces required likely would also have reached MTW-like proportions. Furthermore, operations over the Federal Republic of Yugoslavia (FRY) were not a response to "cross-border aggression" in the classic sense. While large-scale, cross-border aggression remains a real threat to our interests, it is not entirely useful as a measure of requirements. There is room in our strategic vernacular for the simple term "war," to describe operations that require significant levels of national resources and will.

Similarly, the phrase "Smaller-Scale Contingency" (SSC) is equally unhelpful. The air war over the FRY was, by definition, an SSC. Yet it required all of the resources of a major war. In another case, the United States forces that entered Bosnia in December 1995, at the time an SSC, consisted of more than 25,000 soldiers. Clearly "small" is relative. Also, the number of service members directly involved in an operation at any given point does not adequately describe the full resource commitment, which is frequently 3-5 times those actually deployed. Experience, such as that in the Balkans, also tells us that SSCs can evolve into long-term stability operations that increase the stress on our forces and affect our overall strategic flexibility. In particular, there are significant challenges and limitations in our ability to redirect "committed" forces to other tasks around the globe.

This illuminates the risk, mentioned in the NSS, associated with fulfilling our two MTW requirement from a posture of engagement—shorthand for the requirement to withdraw from all contingency operations in the case of two wars. First, the political and strategic consequences may be too high to execute such a withdrawal when the time comes. Second, there is no guarantee that the necessary political decisions would be made in time. Most importantly, no commander is willing to put America's sons and daughters in harm's way untrained. Thus, even with an early political decision, the time required to withdraw, retrain, and redeploy forces to war may well not meet our strategic requirements. While some would argue for establishing specialized forces for peacekeeping, this would not solve the risk challenge, but exacerbate it. As units would have to rotate deployments for overseas peacekeeping, this would require a dramatic increase in end strength, or severely overtax an already stretched force. In addition to the complexities of creating a two-tier force, there is compelling evidence that combat trained troops make the best choice for operations in ambiguous environments characterized by "no peace, no war."[3]

There are other sources of risk as well. The current security strategy also makes no explicit provision of forces for Homeland Security (HLS). While many believe a direct attack on America, by any number of means, is only a matter of time, the military capabilities that would be required to respond to such an attack—or perhaps even multiple, simultaneous attacks—are not accounted for in our current strategy or force-sizing standard. Strategic and theater sea and airlift are also a concern. Recent studies have established that we have a two MTW requirement but only one MTW's-worth of lift.[4] Moreover, this shortfall is based only on a short-term assessment and does not take into account lift needed to meet the capabilities sought in future joint operational concepts and forces. Finally, the need to be prepared to fight two nearly simultaneous MTWs and meet current operational obligations, such as military engagement and stability and support operations, places significant strain on the Services' ability to transform. Real transformation, which is defined as "a fundamental change in our military capabilities," will initially increase risk to our security. Transforming forces are typically unavailable for current operations. The Army is taking some risk now, for example, with the transformation of its units at Forts Hood and Lewis.

Taken together, the problems inherent in our current strategy and force-sizing paradigm stem mainly from the undeniable fact that we have allowed available resources, rather than strategic requirements, to determine our military capabilities. The corrosive effect of the chronic strategy-capability mismatch has put the military on a glide path to unacceptable risk. This risk derives from multiple sources and is the difference between our stated strategic and military ends, and the ways and means available to achieve them. As we engage in this new security debate, we must consider carefully the full range of capabilities required for comprehensive security in an evolving strategic and operational environment.

A Strategy of "Comprehensive Security."

> . . . for years to come Americans will be increasingly less secure, and much less secure than they now believe themselves to be.
>
> Hart-Rudman Commission,
> *New World Coming*, September 1999

One's view of the world and America's role in it is the most fundamental determinant of one's beliefs about a security strategy for the new century. A view that best serves our nation's interests would be balanced. It would be neither excessively optimistic nor pessimistic. It would not hold that America could any more fix all of the world's problems than it can turn its back on them. It would have at its core an equally balanced view of the nature of mankind—that, even in a changing world, the human capacity for both good and evil continues. It would acknowledge that, ultimately, intercourse between states and other international actors is largely driven by the quest for power, where power is measured largely in terms of wealth, and the possession of which requires control over people, land, and resources. It is the quest and struggle for power that may lead to war, and in matters of conflict and men, there are no easy, cheap, or high technology solutions. The nature of war endures, even as the means with which to wage it change. Despite the continuing likelihood of war and lesser conflict, however, America is well poised to continue to prosper, even thrive, in this world. There are as many opportunities as challenges.

While we would like to establish definitively how and when we will employ our power, particularly military power, this goal will remain elusive. Balance, good judgment, and recognition of our enduring interests and values are required. Even though the line between strategic overreach and lost opportunity is difficult to mark, an appropriate national strategy would recognize our enduring national interests and responsibilities, as well as the multiple, complex challenges to them. It would acknowledge the fundamental role peace and stability play in both our own and global prosperity. It would accept the imperative of achieving new levels of synergy among the diplomatic, economic, military, and informational instruments of power. This strategy would recognize that, ultimately, the military instrument, while not sufficient, is necessary for prosperity. It would be informed by both our successes and failures over the past decade and evolve accordingly. Most importantly, it would be comprehensive in its approach, recognizing the essential interrelationship between Shaping, Responding, and Preparing, and synthesizing these elements to achieve unity of purpose. In a world of increasing knowledge and speed, a strategy of **Comprehensive Security** would be increasingly precise, flexible, and deft in its application. It would be far less science and much more art—where *Shape, Respond,* and *Prepare* would be better termed ***Engagement***, ***Preeminence***, and ***Transformation***.

Engagement entails remaining proactively involved in advancing our global interests and fulfilling our global responsibilities. It is an approach to international relations that, while best executed with an abiding sense of national humility and acknowledging the important interests of others, provides unambiguously for our own most important interests. In contrast to its predecessor "Shaping," it recognizes that even our closest allies do not take well to being "Shapees." Moreover, explicitly establishing it as a critical pillar of Comprehensive Security highlights that, unlike the Clinton administration's national security strategy, *Engagement* is not an end in itself. And, while *Engagement* must be comprehensive, involving the diplomatic, economic, military, and informational instruments of power in ways that artfully reinforce one another, it must be based on a recognition that war will not soon cease to be a means to achieve ones objectives. Armed conflict will remain an enduring feature of the strategic landscape. As such, the United States must maintain the capabilities and capacity to perform the full range of military operations.

Preeminence is required for both deterrence and full spectrum dominance. It is achieved through comprehensive strength over time and across the full range of military operations. It acknowledges the dynamic, relative, multidimensional nature of power, and recognizes that neither diplomatic nor economic superiority can be achieved without military dominance. It holds that deterrence and reassurance are the most fundamental measure of effectiveness of the military instrument of power—that the military need *not* necessarily be used in an ultimate confrontation. The Cold War was not won, however, without the resolve and capabilities demonstrated in various theaters throughout the world over the course of nearly five decades.

It should be clear that any strategy which provides the comprehensive security our nation requires will demand that the military deter and if necessary act as the final arbiter of conflicts. We must be able to respond to and dominate in any mission scenario, protect the homeland, and, without abdicating our concurrent global responsibilities, perform critical shaping or engagement tasks. Simply put, we must be full spectrum capable. And, while dominance across the range of military operations is the goal, recent experience tells us that we have quantitative and qualitative gaps in our current capabilities that have manifested themselves both in day-to-day operations and, more importantly, in our ability to deter with conventional forces.

The Hart-Rudman Commission concluded that, "Deterrence will not work as it once did; in many cases it may not work at all."[5] Some might argue that, in principle, deterrence *does* still work the way it used to—that one will be deterred if the perceived risk of loss is greater than the potential for gain. The Commission's conclusion does, however, point out that the United States lacks the *capabilities* to deter certain adversaries and threats. For example, our nuclear triad and high-tech conventional capabilities did not deter Slobodan Milosevic from his campaign of ethnic cleansing in Kosovo—the United States had no "perceived," politically acceptable capability between diplomacy and bombs over Belgrade. Army transformation will narrow this gap between capabilities and strategic requirements.

Transformation is far more than reorganization; it is a fundamental, quantum improvement in capabilities. Transformation in the ways U.S. national power interacts with other states and with international and non-governmental organizations is essential. Equally necessary, and of even higher priority, is transformation of the ways in which our own instruments of power interact to advance our enduring interests and prepare for possible contingencies both at home and abroad. The increasing complexity of the strategic environment makes it imperative that the political, economic, military, and informational dimensions of power be applied holistically and with clear unity of purpose.

A myriad of transformation initiatives are currently under way. The challenge remains to integrate and synchronize these initiatives into a coherent whole. At a minimum, the multidimensional challenges in the current security environment require transformed, multidimensional capabilities. Moreover, to achieve transformation in a world where many technologies are obsolete in months, we must also change the way we change. We do now have a strategic window of opportunity to transform, but we must get on with it. None of us knows with certainty how long that window will be open.

Landpower's Essential Contributions to Comprehensive Security.

While the Marine Corps provides essential, complementary capabilities, the Army constitutes the preponderance of America's landpower capability. Furthermore, only the Army is capable of *sustained* land dominance. As the nation's decisive force, when employed as part of the joint, multinational, and interagency team, the Army ultimately determines the pace of victory and the character of peace.

A simple crosswalk between the pillars of Comprehensive Security and Army capabilities to fulfill that strategy is illuminating.

The Army and Engagement. The Army has a unique ability to set conditions in an uncertain security environment by addressing the human dimension of international relations and potential conflict. Engagement is far more than presence, it is the day-to-day interaction with friends and potential adversaries that reassures and deters. While many countries do not have substantial air or naval forces, virtually all countries have armies. The Army is uniquely suited to engage these armies to promote regional stability, encourage democratic institutions, and set the conditions for more effective crisis response.

The Army and Preeminence. As part of a joint force, the Army is uniquely capable of sustained, full spectrum land dominance. Central to this contribution to America's defense is the Army's ability to conduct decisive combat operations. This credible, demonstrated capability is the foundation of conventional deterrence and underwrites the Army's effectiveness in all other missions, as well as the effectiveness of other elements of national power such as diplomacy. From its ability to conduct precision fires and maneuver; command joint and multinational formations; conduct logistics operations for the joint force; expand the force in a protracted crisis; and, support civil authorities at home and abroad, the Army's contribution to our security is indispensable. This role will only expand as requirements for Homeland Security and missile defense mature, missions in which the Army inevitably will play a central part.

The Army and Transformation. Americans expect their Army to prepare for future conflict, not the last war. In short, they expect their Army to have a vision. The Army does. That vision is to remain *Persuasive in Peace, Invincible in War* in the new century. The transition of the Army, from a Cold-War force to one optimized for comprehensive security requirements of a new century, is imperative, urgent, and under way. Throughout its transformation, the Army must retain the capabilities to fulfill its nonnegotiable contract with the American people—to fight and win the nation's wars. The choice between near- and long-term readiness is a false one.

In particular, a transformed Army's ability to put a combat-capable brigade on the ground anywhere in the world in 96 hours, followed by the balance of a division a day later, is a capability that begins to fill the current gap in conventional deterrence and, if deterrence fails, offers the NCA the opportunity for strategic preclusion. The ability to deploy five divisions anywhere in the world in 30 days ensures that our nation will be able to achieve its political and military objectives, rapidly and decisively.

Force Sizing for Comprehensive Security.

A strategy of Comprehensive Security would have important implications for the armed forces that implement it. In particular, the two MTW force-sizing paradigm would be replaced by a Core Military Capability (CMC) for our operational forces which would have a minimum of four components.

First, the CMC would retain the requirement to fight and win two wars, worldwide, in overlapping timeframes. As the Chairman of the Joint Chiefs of Staff noted recently:

> As a global power, I submit that we cannot retreat from any one activity to do another exclusively. And in this regard, a two Major Theater War capability—and let me stress that this is a capability, not a strategy—has served us well. A two MTW capability allows us to go in two directions at one time. It helps define us as a global power. And when we're committed in one area of the world, it deters opportunistic aggression in another; it helps guard against being surprised. Moreover, it strengthens our coalition relationships—so important to how we will fight in future operations.[6]

One reason for the shortcoming in the two MTW force-sizing construct is that it was designed as a simple tool to address one of several possible "Respond" contingencies, albeit the one perceived at the time as the most dangerous. Although fighting and winning wars is nonnegotiable, Comprehensive Security also demands that America's military continues to do far more than that. As the past decade reminds us, we may be called on to do everything from fighting wars to fighting fires. While this is not new, the Cold War may have obscured this fact. Therefore, if Comprehensive Security is to overcome the deficiencies of the current "Engagement" strategy and two MTW force-sizing standard, it must also recognize that, while a two-war capability is imperative to sustain our status as a global power, it is not sufficient.

This leads to the second component of America's Core Military Capability. We can no longer treat all other missions as "lesser included." As mentioned above, rather than sizing our armed forces to meet current and future tasks, today's strategy addresses quantitative deficiencies by insisting unrealistically that our forces employed in long-term stability operations—such as in the Balkans, the Sinai, Latin America, and other less well-known locales—withdraw, retrain, and redeploy to a second fight. America's CMC must recognize and include the capabilities to sustain essential engagement and stability and support operations.

Third, we must have set-aside forces to ensure the security of the homeland. There is broad consensus on the growing threat array of ways and means by which an adversary, state or nonstate, might directly attack the United States. Under current plans, the capabilities and forces required to defend America are also programmed to deploy to fight the nation's wars, thereby exposing our homeland even further. Clearly, this is unacceptable and must be rectified under a strategy of Comprehensive Security. America requires the ability to defend itself regardless of whatever other contingencies it might be involved in around the world.

Finally, America's Core Military Capability must take into account transformation. While "Transformation" ultimately addresses qualitative shortcomings so that the military becomes a force optimally designed to dominate across the spectrum of operations, we cannot deny that as we transform, there are monetary, force-sizing, and near-term readiness costs. This must be accounted for in a strategy of Comprehensive Security.

The sum of these four "components" of America's Core Military Capability—along with the institutional capacity necessary to field and train the world's premier armed forces—is the minimum essential required to provide America the Comprehensive Security we require. The need to maintain world-class institutional capabilities cannot be taken for granted—the tip of the spear is no sharper than the forces that generate, project, and sustain it.

Conclusion.

Inevitably, the QDR will stimulate spirited debate and demand tough choices. Our nation has enduring global interests and responsibilities, and the threats to those interests are complex, uncertain, and expanding. Our nation requires a strategy of Comprehensive Security supported by a Core Military Capability.

The problems inherent in our current strategy and force-sizing paradigm derive largely from a chronic strategy-capability mismatch and have put the military on a glide path to unacceptable risk. This trend can and must be reversed. As we engage in this new security debate, we must consider carefully the full range of capabilities required for comprehensive security in an evolving strategic and operational environment. Our global interests and responsibilities demand a two-war capability. Additional capabilities are required to carry out critical engagement, stability and support missions, homeland security, and essential transformation tasks. All of these strategic requirements must be met to fulfill our Comprehensive Security needs and mitigate operational, strategic, and ultimately political risk over time. And, as we make the tough resource decisions, the soundest investments are in those forces and systems that have full spectrum effectiveness.

ENDNOTES - CHAPTER 10

1. *A National Security Strategy for a New Century*, The White House, December 1999, p. 19.

2. *Ibid.*, p. 19.

3. The performance of U.N. troops in Bosnia prior to Dayton, as compared to the performance of combat formations under NATO after, is but one clear example of the need for combat capable forces in many peace operations.

4. Estimates of U.S. military airlift shortages vary from 17 to 30 percent of the capacity required to fight two major wars. The discrepancy between these estimates, the former by CINC-USTRANCOM and the latter by the GAO, is due to differences in counting planes—TRANSCOM uses assigned strength, while the GAO decrements that number based on maintenance nonavailability and other usage factors. In either case, both TRANSCOM and the GAO would agree that the result is increased risk, in terms of U.S. casualties, in the event of two wars. For more, see Christian Lowe, "Military Not Able to Meet Requirement for War," *Defense Week*, December 18, 2000, p.1.

5. *New World Coming: American Security in the 21st Century*, The United States Commission on National Security/21st Century, September 15, 1999, p. 8.

6. General Henry H. Shelton, "The National Military Strategy and Joint Vision 2020," *Army Magazine*, January 2001, p. 7.

ABOUT THE EDITOR AND AUTHORS

STEVEN METZ became Acting Chairman, Strategy and Planning Department, in September 2000. He has been Research Professor of National Security Affairs in the Strategic Studies Institute (SSI) since 1993. Dr. Metz also has held the Henry L. Stimson Chair of Military Studies at the U.S. Army War College. Prior to that, he served on the faculty of the Air War College, the U.S. Army Command and General Staff College, and several universities. He has also served as an advisor to U.S. political organizations and campaigns, testified in the U.S. Senate and House of Representatives, and spoken on military and security issues around the world. Dr. Metz is author of more than 80 articles, essays, and book chapters on such topics as nuclear war, insurgency, U.S. policy in the Third World, military strategy, South African security policy, and United Nations peace operations. He holds a B.A. in Philosophy and a M.A. in International Studies from the University of South Carolina, and a Ph.D. in Political Science from the Johns Hopkins University.

WALTER NEAL ANDERSON, a colonel in the U.S. Army, is Chief, Strategic Plans, Concepts, and Doctrine Division, Strategic Plans and Policy Directorate, Office of the Deputy Chief of Staff for Operations and Plans, Headquarters, Department of the Army. Colonel Anderson was commissioned as an Armor Officer in 1978 upon graduation from the United States Military Academy. He has served in a variety of command and staff positions in the United States, Asia, and Europe. From June 1995 to June 1997, Colonel Anderson commanded the 1st Battalion, 37th Armored Regiment, which deployed to Bosnia for a year as part of NATO's Implementation Force during Operation *Joint Endeavor*. More recently, he was a Senior Service College Fellow at the Atlantic Council of the United States.

MICHAEL CASEY currently serves as U.S. Representative Vic Snyder's (D-AR) Legislative Director. He has worked for Representative Snyder for the past 4 years. Prior to that, he served on the staff of U.S. Representative Carrie Meek (D-FL) for 2 years. He holds B.A.s in Political Science and History from the University of Kentucky and an M.A. in National Security Studies from Georgetown University.

DANIEL GOURE is Deputy Director of the International Security Program at the Center for Strategic and International Studies (CSIS). He joined CSIS in 1993 following 2 years in the Office of the Secretary of Defense as Director of the Office of Strategic Competitiveness. He began his career with the U.S. Arms Control and Disarmament Agency and specialized in regional arms control issues. He has been a defense analyst with System Planning Corporation, R&D Associates, Science Applications International Corporation, and SRS Technologies. Dr. Goure has published widely in such journals as *Orbis*, *Comparative Strategy*, *Military Technology*, and *NATO's 16 Nations*. He has been an adjunct professor at Johns Hopkins University and a lecturer at the National War College, Air War College, Naval War College, and the Naval Postgraduate School. He holds an M.A. and a Ph.D. from Johns Hopkins University and a B.A. from Pomona College.

ANTULIO J. ECHEVARRIA II, a U.S. Army lieutenant colonel, is currently assigned as the Director of National Security Affairs in the Strategic Studies Institute. He graduated from the U.S. Military Academy in 1981, was commissioned as an armor officer, and has held a variety of command and staff assignments in Germany and Continental United States; he has also served as an assistant professor of European history at the U.S. Military Academy; as the Squadron S3 (Operations Officer) of 3/16 Cavalry; as Chief of BN/TF and Bde Doctrine at the U.S. Army Armor Center at Fort Knox; as an action officer at the Army After Next project at HQ TRADOC, Fort Monroe, VA; and as a speechwriter for the U.S. Army Chief of Staff.

Lieutenant Colonel Echevarria has published a book, *After Clausewitz: German Military Thinkers before the Great War*, with the University Press of Kansas (2000), and articles in a number of scholarly and professional military journals to include *The Journal of Strategic Studies*, *The Journal of Military History*, *War in History*, *War & Society*, *Parameters*, *Joint Force Quarterly*, *Military Review*, and *Airpower Journal*. He is a graduate of the U.S. Army's Command and General Staff College, and holds M.A. and Ph.D. degrees in history from Princeton University.

RICHARD L. KUGLER is Distinguished Research Professor at the Institute for National Strategic Studies at the National Defense University, where he advises senior DoD officials on national security policy, defense strategy, and force planning. Currently he is leading a major research effort on globalization's impact on national security. Prior to joining INSS, he was a Research Leader and Senior Social Scientist at RAND during 1988-1997. Before that, he was Director of DoD's Strategic Concepts Development Center and a Senior Executive in the Office of Secretary of Defense (Program Analysis and Evaluation). Dr. Kugler is author of multiple books and many DoD studies, and his articles have appeared in *Foreign Affairs*, *Survival*, and other journals. He is coeditor of two books on globalization and national security that will be published in 2001. He holds a Ph.D. from the Massachusetts Institute of Technology, and is both a political scientist and operations research analyst.

MICHAEL E. O'HANLON is a Senior Fellow in the foreign policy studies program at the Brookings Institution where he specializes in U.S. defense strategy and the defense budget, military technology, Northeast Asian security, and humanitarian intervention. He is also Adjunct Professor at the School of International and Public Affairs at Columbia University, and a member of the International Institute for Strategic Studies as well as the Council on Foreign Relations. Dr. O'Hanlon's most recent Brookings book is *Technological Change and the Future of Warfare* (copyright

2000), which describes and assesses the hypothesis that a contemporary revolution in military affairs is within reach. He has also recently completed a coauthored book with Ivo Daalder on NATO's recent crisis and conflict with Slobodan Milosevic, entitled *Winning Ugly: NATO's War to Save Kosovo* (also Brookings, 2000). Dr. O'Hanlon is presently completing two Brookings books for publication in 2001, one on defense choices for the Bush administration and a second with James Lindsay on the issue of national missile defense. He holds an A.B., M.S.E., M.A. and Ph.D. from Princeton University.

IAN ROXBOROUGH is Professor of History and Sociology at the State University of New York at Stony Brook. He was previously employed at the London School of Economics and the University of Glasgow. During 1997-98 he was a Senior Visiting Fellow at the Institute for National Strategic Studies at the National Defense University. Dr. Roxborough has published several books and numerous articles on Latin America. His current research is a study of the post-Cold War debates about U.S. military strategy and doctrine. Dr. Roxborough has a B.A. from the University of York, England, and a Ph.D. from the University of Wisconsin-Madison.

ROBERT D. STEELE is the author of *On Intelligence: Spies and Secrecy in an Open World* (AFCEA International Press, 2000), and the founder and CEO of Open Source Solutions Inc. (OSS). He and his company are featured in *The Year in Computers* (2000), he has been twice named to the *Microtimes* 100 list of "industry leaders and unsung heroes who . . . helped create the future," and is featured in the chapter on "The Future of the Spy" in Alvin and Heidi Toffler's *War and Anti-War: Survival at the Dawn of the 21st Century*, among other publications. He has written widely in the areas of information strategy, open source intelligence, and asymmetric warfare including information warfare. In the course of a 25-year national security career, Mr. Steele has served as a Marine Corps infantry officer and service-level plans officer; fulfilled clandestine, covert action, and technical collection duties; programmed funds for overhead reconnaissance capabilities; contributed to strategic signals intelligence operations; managed an offensive counterintelligence program; initiated an advanced information technology project; and been the senior civilian responsible for founding a new national intelligence production facility. Mr. Steele holds graduate degrees in international relations (Lehigh University) and public administration (University of Oklahoma), and certificates in intelligence policy (Harvard University) and defense studies (Naval War College).

JOHN F. TROXELL, a U.S. Army colonel, is currently serving as the Director of National Security Studies, Department of National Security and Strategy, U.S. Army War College. He has held assignments in the Department of Army War Plans Division, as a force planner for the Assistant Secretary of Defense for Strategy and Requirements, and was a former economics instructor at the U.S. Military Academy. Colonel Troxell's operational assignments have included service with the 1st Infantry Division, 293rd Engineer Battalion, and the 43rd Engineer Battalion. He also commanded the 3rd Engineer Battalion, 24th Infantry Division. His most recent assignment prior to joining the faculty at the War College was as Chief, Engineer Plans Division, Combined Forces Command, Seoul, South Korea. Colonel Troxell has published articles in *Parameters, Military* Review, and his previous publication from SSI was *Force Planning in an Era of Uncertainty: Two MRCs as a Force Sizing Framework.* He earned a Bachelor's degree from the U.S. Military Academy and a Master's degree from the Woodrow Wilson School, Princeton University. He is a 1997 graduate of the U.S. Army War College.

HUBA WASS DE CZEGE, a retired U.S. Army brigadier general, was a career infantry officer with 2 years combat experience at the small unit level in Vietnam, and command and staff at all levels up to assistant division commander at retirement in December 1993. He was one of the principal developers of the U.S. Army's *Desert Storm* and *Just Cause*-winning operational concept, AirLand Battle. Brigadier General Wass de Czege was the founder and first director of the U.S. Army's premier graduate school for military planners, the School of Advanced Military Studies (SAMS) at Fort Leavenworth, KS. During NATO's crucial transformational years, from 1988-91, he was an advisor to SACEUR, General "Jack" Galvin, and NATO Secretary General Manfred Woerner on military strategy relating to arms control and post Conventional Forces in Europe (CFE) treaty strategy. From 1994 to 1996 he taught Applied Military Strategy at SAMS. In 1994 he began studying the effects of new technologies on the future practice of the military art. During the last 4 years, he has been fully engaged in the Army's Army After Next studies project, and various other DoD consultancy assignments to include service on a Defense Science Board summer study panel (1998) and on several Defense Advanced Research Programs Agency (DARPA) advisory panels. Brigadier General Wass de Czege was educated at West Point (USMA, 1964) and Harvard (MPA, 1972).